Scanning and Sizing the Universe and Everything in It

Scanning and Sizing the Universe and Everything in It:

Playing Scales

By

Harold Toliver

Cambridge Scholars Publishing

Scanning and Sizing the Universe and Everything in It: Playing Scales

By Harold Toliver

This book first published 2023

Cambridge Scholars Publishing

Lady Stephenson Library, Newcastle upon Tyne, NE6 2PA, UK

British Library Cataloguing in Publication Data
A catalogue record for this book is available from the British Library

ISBN (10): 1-5275-9236-7
ISBN (13): 978-1-5275-9236-0

From this the poem springs: that we live in a place That is not our own and, much more, not ourselves And hard it is in spite of blazoned days.
(Wallace Stevens, "Notes toward a Supreme Fiction")

< >

That the heavy elements in our bodies directly connect us to cataclysmic events which occurred billions of years ago in our galaxy is extraordinary enough. But this is only part of the story. For the heavy elements in our bodies may also connect us to a time so impossibly remote that at present we have only the vaguest knowledge of it. That time was before the birth of the Milky Way itself. (Marcus Chown, *The Magic Furnace*, 211)

< >

TABLE OF CONTENTS

PREFACE

The scalar spectrum, from absurdly small to unimaginably large, is too extended and counter-intuitive to be easily assimilated. Because our genetic ancestry developed in response to a narrow range of what exists in totality, we get no direct sensory reports on anything at the extremes – in high velocities, brief and extended time spans, lengths, distances, in X-rays and gamma rays, even in high-pitch sounds. The 'gut feeling' that ancestry passed along serves some purposes quite well, making us wary of precipitous slopes and animals with fangs, but it is of no use in understanding what has never been part of anyone's DNA. For that, we need magnification and sensitive detection equipment in the hands of experts, and numbers given awkward names like vigintillion and centillion. Extremes do not come naturally to us - and nothing that does come naturally can be assigned its appropriate place in an overall scheme of things without our knowing something of the rest. Natural philosophy is dedicated to keeping that in mind - placing *each* in the context of *every*. To safeguard our understanding of these issues against human-centric beliefs, we need to be aware of that – even where we don't have exact figures for distances, times, temperatures, and only estimated numbers for galaxies and stars.

Measurements are one area of interest in our adjustments to variable scales, relativity another, and nature's heterogeneity, its sheer profusion, a third. Over time, the latter has generated stars in the sextillions (a wild approximation) and galaxies in the hundred-billions. It has created millions of living species, areas of celestial wreckage light-years wide, put disks of debris around proto stars, generated nebular clouds of dust and gas larger than the solar system, collided large chunks of matter, created tadpole tails of matter behind stars under spatial storms, tossed out energetic oceans of radiation, and exploded supernovas in unimaginable sizes - much of these discrediting venerable beliefs around the globe. As I argued in *Mythic Worlds and the One you Can Believe In* (2017), natural philosophy advises strongly against adopting any cultural framework incompatible with that extended reality, from quarks to quasars. World views that propose a master controller are especially prone to ignoring nature's harshness, extended areas of chaos, and uninhabitable extremes. Hard it is indeed in spite of blazoned days as Stevens writes and everyone knows.

In his "compact history of infinity," David Foster Wallace lists a few things - now familiar - that still look odd from our customary perspective, but have become commonly acknowledged by science watchers: "time is relative," we've been told many times, meaning relative to changes in motion and speed. "Quantum particles can be both there and not. . . space is curved. . . colors do not inhere in objects themselves . . . astronomic singularities have infinite density our thoughts and feelings are just chemical transfers in 2.8 pounds of electrified pâté. We 'know' a near-infinity of truths that contradict our immediate commonsense experience of the world. And yet we have to live and function in the world" (22). Add to these the spooky action at a distance of two bodies that attracted Newton's attention. The subatomic level is as troubling as the expanses, the extremes in size and distance, and the range of temperatures from -273.15° Celsius or -459.67° Fahrenheit to billions of degrees. It doesn't take much to send some particles spinning so far out of control that only cooling to near absolute zero will quiet them down. That, too, is an extreme never seen outdoors on our favorite planet, only under lab conditions.

Something not quite that wild is true of the bands of debris that collect around stars, three around the sun counting a fuzzy Oort cloud in the far distance.

Some of the better insights into, and applications of, the scalar spectrum can be found in essays devoted to nearby nature. This is especially true of those concerning the American West, where much of the geography in deserts, deep canyons, and towering peaks is in a wild, unruly state. That wildness is indicative of what has come within the range of magnified vision. In Reg Saner's "Wind" in *The Four-Cornered Falcon* (1993), the force of an untamed wind makes the author mindful of levels at work within him down to molecules. Upon dying, he realizes, "not one of my personal molecules will ever again take anything personally. The pure animal truth of being wiped utterly out. That terrifies me." (Subjectivity breaks in.) "Yet I feel the rightness of that, as part of what is. So be it" (54). (Objectivity answers.) That is less difficult to follow for someone who has juggled objectivity and subjectivity without fumbling too much, which is basically what learning to play scales amounts to. It is an attitude and an art, as well as a necessary branch of any comprehensive account of reality.

A plangent elegiac note is likely to creep in when longer and briefer spans are the subject, a source of much philosophy. Mortality, among atomic matter's cycles of buildup/breakdown/- buildup, is the biological branch of atomic matter's energetic turnover, clustering in buildup, scattering in breakdown. The universe itself seems to be doing that, sending its galaxies racing away from the tighter configuration of the distant past.

Getting the cycles and what they produce as accurately gauged as possible falls mainly to the sciences - the macrocosm to astrophysics, astronomy, and cosmology; the microcosm to physics and chemistry. Applying these to common assumptions falls to natural philosophy, a term that, for lack of a better one, I borrow from both the ancient Greeks and from general use in place of *science* until the 17th century. Distinguish that branch of serious thinking from moral and aesthetic philosophy. To get *each* properly in the company of *every* requires at least a passing acquaintance with some specialized areas of research. No better way exists to protect against devoutly believing in one of the regional group-think convictions, which continue to raise trouble around the world.

Since getting the hang of extended times, masses, temperatures, sizes, and distances is a lifelong endeavor, I've marked selected stages of doing so, of scale playing, with a model case seen partly from the inside, as a human template. What does it actually feel like to make adjustments to reality, one by one, as we get better acquainted with it? We won't individually take exactly the course McLaurin (as I'll call him) does, but something close enough to it to use his advance to establish way-markers. Even childhood – *especially* childhood, in fact – is pertinent, because everyone takes approximately the same sensory equipment out to play. It is only natural to think that what we see is what is, instead of just one level of it. McLaurin will eventually review his earlier impressions in the context of the natural continuum. Like everyone, he notices early on the contrast between the biosphere's bristling profusion and organized human areas, fields, gardens, households - some of which I've loaned him from my own storage. On its logistic or bifurcation charts, chaos theory now captures some of nature's less orderly and shapely production of objects, detritus, and clouds of dust and gas trillions of miles across. In due course, McLaurin will come to their charts and to nature's proliferation.

No one can be master of many disciplines, or reconstruct anything like the entire natural continuum. Even so, most can learn to see more clearly how one thing has led to another in deterministic succession. As Alexander Humboldt and John Muir realized two centuries ago, "when we try to pick out anything by itself, we find it hitched to everything else in the universe." More recently, Edward O. Wilson in *Consilience* (1998) has endorsed that principle, just as the following chapters do. My hope is that a good many others will do something similar - and soon, before the narrow range of living conditions shrinks even more. The planet's air and water temperature, its population, its supply of fresh water, and its food are in critical condition.

Misconceptions and quarrelsome beliefs contrary to one another, and to natural history overall, hinder the collective effort necessary to sustain a such matters. Story value guaranteed no correspondence to reality, however. The sun is always there whether or not a given culture has a sun god. The natural continuum has followed a deterministic course in producing the atomic matter succession. It is, after all, atoms all the way down, and out and up, and atoms don't have minds of their own. They do exactly as their unswerving laws dictate. As far as we know, in our section of the universe, nothing deviated so much as a charmed quark from necessity. That is, until a few million years ago, when animals started making choices within the narrow range allowed by biology and circumstance.

One reason for the persistence of people-oriented explanations despite abundant evidence to the contrary is that, until the 20th century, not much was known about what was too small or distant to see. Another problem proved troublesome as well: scientists and philosophers from the time of the pre-Socratic philosopher Thales of Miletus (c.546) and Democritus (5th century BCE) onward believed on logical, not observational, grounds that at the small end a minimal level had to exist. Otherwise, motion across infinitely divisible things and distances would be impossible, the reasons for which I'll return to in a moment (with respect to Parmenides and Zeno). What were prematurely named *atoms* ('uncuttable' units) didn't actually become visible until the invention of electron microscopes, and weren't the smallest units, since they had protons that in turn had quarks - a term the physicist Gell-Mann took from James Joyce. *Atom* had been used by Democritus and Leucippus to describe the smallest bits at the grainy bottom level of matter.

In the opposite direction, telescopes in the 17th century, which presently enable astronomers to see outward and backward into spacetime to nearly the beginning of the universe, began making things visible that weren't supposed to be there. The gap between science and common beliefs widened considerably with Copernicus (1473-1543), Tycho Brahe (1646-1601), Galileo (1564-1642) – equipped with the first telescope – and Johannes Kepler (1571-1630). The beginning of the microscopic phase came soon after, when Anton van Leewenhoek (1632-1723) magnified microbes that he called *animalcules,* squirming little things that fascinated members of the Royal Society as well. Mechanical atomists in chemistry paralleled those discoveries in the realm of the very small. Daniel Sennert, Robert Boyle, Pierre Gassendi, and Isaac Newton followed up on Epicurus in pursuit of the building block level that, heretofore, had everything being composed of just four elements: earth, air, fire, and water.

Those who followed science in the 18th through the 20th centuries continued the tug of war between objectivity and various versions of a people-oriented universe. In the mixed philosophic and scientific era of Hobbes, Descartes, Gassendi, Boyle, Newton, and Locke, the renewal of atomism and empiricism didn't yet propose a purely natural or materialistic continuum. Instead, philosophy and science were squeezed between observation and traditions, upheld by influential institutions. Prevailing beliefs discouraged discoveries that didn't agree with them. Despite what they were seeing, the users of microscopes and telescopes assumed that a master design was manifest in nature – in stars, volcanoes, mountains, tornadoes, predation, and the rest.

The topic of the moment isn't that history of oddities and errors, however, but the gap between the human range of sensory impressions and the stretched scalar spectrum. None of our genetic ancestry needed to see microwaves or feel microbes, and so we don't. While revered traditions attract the finest adjectives and inspire magnificent art and architecture, atomic matter fills blackboards with equations and a nomenclature of fermions and bosons. Getting inventive with what they call super partners, physicists add squarks, gluinos, sgluons, sleptons, gravitinos, and for good measure (pun unavoidable) zinos and winos. Adjusting to what falls outside our normal range is difficult for everyone, average or brilliant, scientific or literary. Having the explanatory mathematics helps on paper, but not much with expanded and shrunken times, very short and very immense distances, and sizes from sub atomic to galactic. No one can feel at home with a hundred billion galaxies. Even Jupiter's 79 (so-far discovered) moons are difficult to take in for inhabitants of a planet with just one.

In ascending orders of magnitude in distance, the scales go from millimeters to kilometers to the astronomic unit AU (distance from the earth to the sun), to light years (5.8 trillion miles), to a favorite of astronomers, cosmologists, and astrophysicists: parsecs at 3.26 light years or 19 trillion miles. Where necessary, they can proceed from there to megaparsecs and even giga- parsecs. The astrophysicist Joshua Winn puts the range of sizes and distances in a convenient table with 1 meter setting the human norm, 10^{-6}m the visible light wave minimum, stars at 7.0 x 10^{8}m, galaxies at 3 x 10^{19}m, and the entire visible universe at 3 x 10^{25}m and spreading. To apply the scalar spectrum and natural continuum as a litmus test to venerable beliefs, we don't need to memorize these and other measurements - we merely need to acknowledge the range. What they gauge is real. The measurements are available on charts and diagrams, in the calculations natural philosophy uses when it is being comprehensive, in narratives of space travel, and in long-range natural history. Measuring the projected future of the universe in powers of ten in seconds, Gerard Hooft and Stefan Vandoren (2014) adopt Jules Henri Poincaré's calculation for a final lapse into darkness at 10^{1000}seconds. Don't bother lining *that* up with anything. Again, merely be aware that reality, in total, is simultaneously minuscule and huge, rapid and prolonged, not beyond measure but beyond human comprehension except in the abstract and in lumpsum. Any conviction or belief incompatible with those measurements is on shaky ground.

To return to Democritus and the atom: Parmenides and Zeno insisted on the logical problem for science, philosophy, and common sense where we keep going smaller and smaller. The quandary, as Zeno posed it, was that any span of time or any distance across a field could be halved, halved again, and then again, toward but never reaching infinity. That made motion through either time or space impossible. As minimal units, atoms – proposed partly in response to the need to shut down the dividing and multiplying – weren't completely satisfactory. One could continue halving even atomic minima, and they, in turn, would have infinitely divisible spaces between them. In leaping from one to another, something trying to move across the gap, would still fall into the Parmenides/Zeno infinity abyss. Physics hasn't yet found a sure-fire escape from that dilemma unless with loop quantum gravity (LQG) (introduced as a possibility recently), which proposes what astrophysicists describe as a spin network or spin foam. Its quanta, on the order of a Planck (minimal) length, are said actually to *compose*, not just *occupy*, space. As advanced by Carlo Rovelli and Lee Smolin among others, LQG thus combines gravity, quantum mechanics, and spacetime in quanta, with loops that connect one to another. Nothing is settled about that, but indivisible quanta are, as usual, made the basic building blocks, and if no space exists between them, all the better. Thought can keep halving all it wants, but they aren't, themselves, divisible - nor can the space between them be if nothing of the kind exists. Then, it becomes loop quanta all the way up, down, and out. What look like spaces, in that case, would become continuous fields - like that of gravity which, likewise. has no gaps to be crossed.

In any case, my concern isn't with these (or any other) philosophic conundrums, but with the implications of well-established measurements. Setting Parmenides and Zeno aside still leaves dark matter and dark energy unsettled, along with what propelled the big bang and inflation at the outset. Add to these the possibility of something smaller than quarks, as yet undiscovered, and the possibility of a multiverse that plays by different rules. The latter comes in hypothetical variants that Brian Greene (2011) names quilted, inflationary, brane, cyclic, landscape, quantum, holographic, simulated, and ultimate (355). We don't need to settle every speculative matter to see what observation, theory, and measurement have established with strong likelihood - practically all of it outside the range of the sensations our genetic ancestry bequeathed us. The scalar spectrum, as well as what telescopes and microscopes reveal, are what we have to utilize in determining the range of what is real, and what the imagination, wish fulfillment, and ego have substituted for it. In online lectures, the astronomer Mark Whittle puts human adaptability to those extravagant measurements in the leading question "how can something that emerges out of the atoms and forces of Nature – ourselves –- ultimately come to understand such a vast and alien realm?" Alien it is indeed, and so vast that we can scarcely use familiar things to gauge it.

Given nature's magnitude and apparent chaos, no one can feel at ease, even with just what the senses report, let alone all the rest.

So, how can we understand such things? One answer is that we can't, not comprehensively, but areas of study in multiple combinations have shown promise of, eventually, accounting not only for dark matter and energy, but for what propels the universe outward. Both how far the dispersing galaxies may have built out distance and how long the exchanges of energy and matter can be expected to last are unknown, but open to speculation. Closer to home, how was it that atomic matter generated life on the planet some 3.5 billion years ago? Chemists and evolutionary biologists already have a reasonably good idea of that. The rock record shows billions of years of single cell life before evolution got inventive in the Cambrian explosion some 540-530 million years ago. The universe was already 5/6th of the way towards the emergence of *Homo sapiens* before those lifeforms came forth, speaking just of this planet, not the billions of other possibilities. (Under the Drake equation named for Frank Drake, the Milky Way alone could have a good many based solely on probabilities. Needless to say, that's highly speculative.)

Despite its missing pieces, the standard model has a good enough pedigree to justify the word *standard.* Here is one common rendering of it. Other versions will follow in due course, but this one is useful for its categorical distinctions among quarks, leptons, the forces, and the Higgs boson:

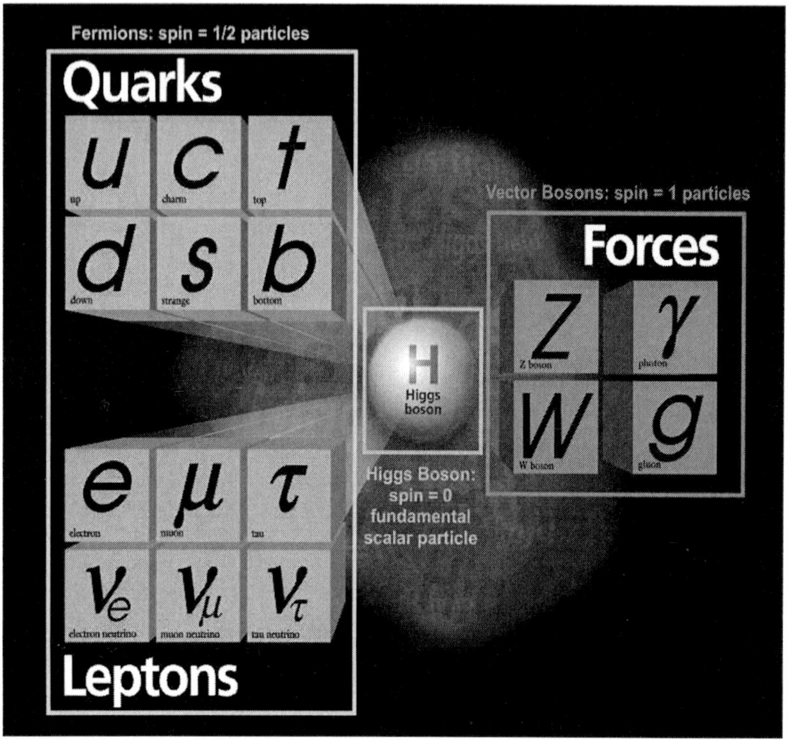

What that standard model compresses into its categories doesn't add up to a theory of everything, because it doesn't combine the nuclear strong force that holds protons together with the electroweak; nor is gravity fully understood at the level of gravitons or the **W**eakly **I**nteractive **M**assive **P**article (WIMP) proposed for dark matter. The phrase *natural continuum*, nonetheless, amounts to an acceptable term of terms, which is how I'll use it. So far as perceptions and observations go, nothing exists outside of it, nor has anything come into succession from beyond it once it got underway an estimated 13.799±0.021 billion years ago, in what is sometimes described as an extremely hot particle plasma. For the projected end of the universe, cosmologists provide only very tentative estimates, but spans of time overall range from the briefest Plank instant to when the Appalachian Mountains will melt down into hills and mounds, the sun will expand into a red giant, and atomic matter will reach a final state of decay or entropy. Somewhere near that farthest extreme, a black hole containing millions of solar masses, decaying by the Stephen Hawking evaporation process, is projected to finish the black hole era in about 10^{106} years.

Something like an ultimate quiescence will come when no more energy or heat passes from one thing to another and nothing further leaks out of black holes.

That degree of speculation passes into science fiction, but the astrophysics on which long range projections backward and forward are based is sound. Nature produces explosions and collisions but no leaps or reversals. In familiar terms, nothing intervened to sink the latest ship that has gone down, or to drown one of its crew and save everyone else - nature did it all, raised the winds, churned the water, gave the crew the lungs that required oxygen, and floated the lifeboat that saved nearly everyone. *Natural history* and *natural continuum* mean the same thing in that respect, with one object or event following another under universal mandates. Measurements enable comparisons, not merely in putting common sensory impressions across from other frames of reference, but in putting any particular frame of reference across from any other and the whole.

The desire to define such a totality has been around since philosophy began. Almost any boundary nomenclature will do for what lies beyond it and beyond measurement; *eternity, nothing*. *God* doesn't serve particularly well because, as animations, deities tamper with natural laws and usually resemble people. Humanizing or deifying what underlies the Carina nebula is an anthropomorphic and illogical thing to do, given its complexity and 230 light year expanse (in miles, about 1,350 trillion). Moreover, if something with the necessary power and planning is credited with creating the gazillion stars, planets, and galaxies, we would also have to hold it responsible for everything that goes wrong - the collisions, breakdown phase of the buildup/- breakdown cycles, and in living things the predation, diseases, pain, and 100% mortality rate.

Personalizing the cosmos can be done only by projecting not merely a supreme power, but an unimaginably malignant one. Nothing with good intentions and adequate power made big fish need to swallow small ones to stay alive.

More about that in due course, but first the sizing problem - when smaller and smaller at one end, plus faster and faster, is set beside bigger and bigger at the other end, plus longer and farther.

Swift's Fleas and Other Tiny Things

Distrust of measurements accompanies them where they dislodge what is considered 'normal', and sometimes divinely ordained. To get an idea of what that meant when the concept of magnification turned up in thc 17th and

18th centuries, – when a comprehensive, feasible natural world was being assembled for the first time – let us look to van Leewenhoek and the Royal Society, and listen to what exceptional mind and talent Jonathan Swift says in response. Frequently in the mix with empiricism, observation, and the development of means to see smaller and farther, was an assumption that what lay out of sight wasn't meant to be pried into. This is still an assumption in some quarters, but is seldom put so well as it was by Swift; he didn't find going small necessarily objectionable in itself, but going too far either up or down raised questions of Gulliver's kind among the Lilliputians. A section of "A Poetic Rhapsody" puts belittling, specifically in poets nipping the behinds of smaller and smaller poets, in nursery rhyme form:

> Big fleas have little fleas, Upon their backs to bite 'em,
> And little fleas have lesser fleas, And so, ad infinitum.

A variant, slightly statelier, moves to a slower cadence, not quite as catchy for the school-going set, but as satire more (mock) dignified:

> So naturalists observe, a flea
> Hath smaller fleas that on him prey,
> And these have smaller fleas to bite 'em. And so proceed *ad infinitum.*

Along with the organization of the poem into couplets and tetrameter, citing naturalists (scientists) is the main difference. Magnifying things is marvelous, but the relativity of the perspective it encourages can be demeaning to the prime species of the universe. As others made explicit, too, the scale of normal observation was set for a reason. When Milton's Eve gets curious about the stars, that's more or less what she is told on the authority of both her husband and the archangel Raphael, sent to fill in what the newly invented humans need to know about what came before Eden. Curiosity beyond that, much like the fruit of the tree of knowledge, is forbidden. For conservative monarchists like Swift and his fellow satirist Alexander Pope, it also carried implications in that those who got too inquisitive could end up questioning the civil and ecclesiastical hierarchies. These after all administered the laws and the sacraments. 'Keep to your assigned place' was implicit alike in the great chain of being, the Eden story, and monarchy.

Swift made prying objectionable on other occasions as well, as did Pope. He hit full stride in that regard in mocking the 'projectors' at the Lagado academy in *Gulliver's Travels*. Lacking faith in providence, a catchall super power that looks after people (except when it doesn't), the academy's astronomers obsess over close calls with disaster: "the earth, by continual approaches of the sun toward it, must in course of time be absorbed or swallowed. The earth narrowly escaped a brush from the tail of the last comet, which would have infallibly reduced it to ashes; and the next, which they have calculated for one and thirty years hence, will probably destroy us" (Voyage to Laputa, chapter two). The mockery implies that such disasters weren't possible, which, of course, they always have been - with or without providence. The first scheduled approach of the sun won't be until its red giant phase, which shan't happen any time soon. That doesn't mean, however, that a sizeable asteroid or the tail of a comet won't crash into the planet.

Disapproval of venturing into forbidden areas was by no means finished in Swift's century, or the next. Science continued to be the butt of satire in Mark Twain's Tumble-Bug, disdainful of Manology (study of the extinct species Man). The species that have inherited the earth, including the snails and insects, agree with Tumble-Bug that "Science only needed a spoonful of supposition to build a mountain of demonstrated fact out of." The moral? Mankind should not be "prying into the August secrets of the Deity" (*Sketches*, 148). Since Twain's day, the "spoonful[s]" have multiplied into truckloads. Another tendency of the 'everything is as it should be' assumption was making suffering out to be for the good of the sufferer. Confidence in Leibniz's best of all possible worlds gets Candide into one scrape after another - though Leibniz didn't actually say the world was perfect, only that God made the most of the available alternatives. The logic was that if in a well-made, infallibly governed universe, nothing can happen that isn't intended, and what looks to be disastrous must not be.

Archived Data

Despite the skepticism over magnifying things, and sometimes over science more broadly, 17th and 18th century naturalists continued compiling quantities of information. As Lorraine Daston and Peter Galison point out in *Objectivity* (2010), 17th century science put together unprecedentedly expansive atlases, bestiaries, and botanical works. After the mid-17th century, publications included studies by members of the Royal Society, one of whom, Robert Hooke (1635-1703), had a microscope of his own. His first publication advertised on its cover page "physiological descriptions of minute bodies," and lived up to its promise with a blown up image of the eye of a fly and an example of a *cell* (a Hooke coinage). A new world of marvels was opening up, this one devoid of miracles but, in the minds of its observers, not without design and intent. Hooke believed that what he saw confirmed a divine plan of some sort. He was prepared to see it that way, and so he did, blocking off what contradicted his and nearly everyone's assumptions. Like it or not, a mosquito is meant to be just what it is; so were animalcules (cells), not yet known to be in league with diseases.

That Hooke's publication was popular didn't mean that it was known to the general public.

At the beginning of the age of enlightenment, reading was still restricted to a fraction of the population, albeit a somewhat larger fraction than a century and a half earlier when Thomas More (1478-1535) taught his daughters to read. Physics landed in its own university department during the lifetime of Newton (1642-1727), where mathematics already had a chair. Crucial to natural philosophy's future, Newton removed the age-old contrast of the celestial to the terrestrial sphere, one perfect and forever stable, the other constantly changing and subject to decay. The same laws applied everywhere; apples fall from trees, and planets are held in orbit around the sun.

The advances in 17th century mathematics, astronomy, physics, and microscopic science were followed by a century of unprecedented investigation, exploration, mapping, and a broadening literate public, contemporary with Swift and Pope and a century after. Georges-Louis Leclerc, Comte de Buffon (1707-1788), collected data on flora and fauna into a 36-volume natural history (*Histoire naturelle, générale et particulière,* 1749–1788). What geology, astronomy, biology, chemistry, and physics were assembling remained only loosely connected, but how far the cooperation of disciplines had come by the mid-19th century is made clear in the naturalist Alexander von Humboldt's (1769 –1859) ambitious work *Cosmos*. Published in its most complete form in 1858, the work was a milestone, putting the natural continuum into strictly mechanical terms. For Humboldt, what linked events and fields of study was a causal chain. He accompanied ventures into the earth sciences with the promotion of that each-in-every and every- in-each principle. The continuum left nothing dangling. As if preparing for a concept of evolution, *Cosmos* presupposed a greatly lengthened earth history. This would support what came the following year in Darwin, whose earlier voyage on the Beagle was accompanied by Humboldt's previous publications. What Humboldt had seen of Darwin's work was minor, but both required a seriously revised world history and an extension of time matching that of space to come several decades later. Connecting planet history to that of the stars was still a long way off, but the foundations were in place.

Humboldt didn't spring up out of nowhere; he had predecessors in earth sciences, particularly James Hutton, a geologist and fellow student of mineralogy. In *A Theory of Earth* (1785) Hutton had found that, because of vertical earth movement, what was on the surface of the earth now hadn't always been. Nothing within the brief scope of the conventional created universe accounted for that. The movement was driven by heat, though the layers had also accumulated by sedimentary means before fusing into metamorphic rock. Clearly, by *The Origin of the Species* (1859), many of the common beliefs about world history had become untenable. Either the presumed powers behind the universe had chosen to work exclusively through nature and be manifest only there, or no such powers existed. Especially prominent in the work of both Humboldt and Darwin was the harsh struggle that forms of life undergo. For Humboldt, South American jungles and mountains supplied abundant examples, as did his subsequent trek across Russia.

Where disciplines were coordinated, the data and theory from each had to be on hand for the others. Data storage and distribution came in high volume in the 17th and 18th centuries. Had Swift been alive post Humboldt and Darwin in the 20th century, he might have ridiculed heaped up facts as well as humans prying into forbidden areas – like this if he were having an off day:

> So naturalists observe, a file Hath smaller bits within its pile, And these have mega mites And so proceed to gigabytes.

Now, on to terabytes. The *tera* prefix (10^{12}, trillion) now comes into play often, as in the teragrams of CO_2 and CH_4 in the atmosphere, light years in kilometers or miles, in the terabyte storage capacity of computers, and the centralized storage of OneDrive and the Cloud. That is no exaggeration. The cosmological theorist Sean Carroll tells us that in the **L**arge **H**adron **C**ollider (LHC) at Cern, particle collisions produce a billion megabytes of data at a rate of a billion collisions a second. What is to be captured and recorded obviously has to be compressed and selected, and sampling is the only workable method.

So numerous has the data become, it requires narrow specialty areas to compile it, and tiers of categories or cascading files to organize it. Induction, as always, goes from incomposite (simple) bits and pieces to elaborate nomenclature pyramids under ever-more inclusive titles, with ever-more bits and pieces and more elaborate file pyramids. Mathematicians carry numbers only so far before resorting to the sign for infinity (00) or ad infinitum for the remainder. Technically, no number can reach there, but, as an absolute, it makes a convenient reference point that means 'and so on, and on, and on'. At the quantum level, calculations of probability run into the quirky behavior of qubits and quarks, these too in lumpsum, since descending in size ascends in multiplication. Ascending by means of abstraction creates packets of convenience. Equations go a step further in compressing much into little. One of them, Erwin Schrödinger's (1925-26) for wave function, covers everything composed of waves. Since Newton's invention of calculus, the scientific tool kit has acquired several - such that, in language, take the form of propositions and theories and in mathematical equations, and visual representations on charts combine with numbers and words.

Whatever the measurement extremes and the details, the philosophical and religious implications of a cosmos governed by natural laws were evident in disturbed post-Darwin commentary. Let us compound that by making it also the post Hutton and Humboldt era, putting Darwin in the context of a growing naturalist movement. The movement had accomplished a good deal before he added the evolution of lifeforms. Attempts to reconcile other notions of origin and universe management continued even among naturalists, somewhat surprisingly in the later essays of John Burroughs.

Let us glance further back again for a moment, this time consulting philosophy rather than, or along with, science. A century after Bacon's apology for induction, Pope argued in "An Essay on Man," "Know then thyself, presume not God to scan":

> The proper study of Mankind is man. Plac'd on this isthmus of a middle state
> A being darkly wise, and rudely great: . . . . Sole judge of Truth, in endless Error hurl'd:
> The glory, jest, and riddle of the World!
> (Epistle II, 1-18)

One key lies in *presume*, the sin of Milton's Eve at the forbidden tree thinking she can rise in stature by a magical infusion of knowledge (not that bad of an assumption, given a talking serpent and the wonders of language that had come to her instinctively). For the conservative Pope, 'presume not' carried the implication I mentioned earlier, 'keep to your place'. In his way of thinking, the word *scan* – applied to an infinity that can't be scanned – exposes the absurdity. Just how seriously we are meant to take that, however, is put in question by his equally noteworthy endorsement of Newton, in God said "Let Newton be! and all was light." The problem lay with science trying to scan no less than the infinite power that sustained the universe.

The next two centuries saw a wealth of observation and theory, from fossil studies to astronomy. This seriously tampered with the Aristotelian,

Platonist, and theological heritage that philosophers, from Hume and Descartes to Locke, were determined to combine with renewed empiricism. Thanks partly to the Watson/Crick discovery of the DNA double helix (1953), the proper study of mankind came to include phenomena like the Hox gene, which tells other genes what to do. The empiricist and naturalist assumption is that nothing is disconnected, from top to bottom and from time 0 to the end of time. Relations among the segments of nature's profusion can be found only by tracing each segment back to a mutual ancestry - going from twigs and small branches of the evolutionary tree to their supporting limbs and on down to the trunk where animal, vegetable, and fungus kingdoms are joined. Complex forms of life came from the single cell (prokaryotic) kind that prevailed for some three billion years. Still further back, multiplication by division was the way prodigious natural production came from the mottled near uniformity of the **C**osmic **M**icrowave **B**ackground (CMB), relic electromagnetic radiation located in 1964 by Arno Penzias and Robert Wilson using radiowave astronomy. No stars or galaxies had yet formed at that early stage, and life on earth was still 9 billion years away, the thinking biped over 13.7 billion years. That was time enough for light to travel the better part of 8×10^{22} or sextillion totally unthinkable miles.

The Symbol User

That's the historical setting for learning to play scales, which has to be done in the company of still-predominant beliefs that are almost as far from the natural continuum and the scalar extremes as Swift and Pope were from the magnification and natural history of their day.

Invention of all sorts put humans at the forefront of the animal kingdom, placing them apart from atomic matter history and the extended scalar spectrum. Imaginary things have always been based on the normal range of human experience. I'll save most of that bundling of things real and fantastic for later, but what we need to keep in mind is it being based on our sensory range. This is an inevitable part of the gap between what's actually out there and how we perceive, represent, and misrepresent it.

The recent flood of contributions as to how language and other means of representation operate gives some indication of the unavoidable discrepancy between words and things. That is underscored in Rene Magritte's surrealistic painting of a pipe labeled "This is not a pipe" - a picture without that double take would have ignored the obvious fact that a two dimensional image on paper isn't a three dimensional object. The contributing cadre on language and communication continues to grow, including in recent decades, especially the 80s and 90s, Jude P. Dougherty (2013), Herbert W. Simons (1990), Michael Hogan (1998), Norman Fairclough (1992), Alan Gross and William Keith (1997), Luc Pauwels (2006), Leah Ceccarelli (2001), Daston and Galison (2010), Daston and Park (1998), W. H. Newton-Smith (1981), George Layoff and Mark Johnson (1980, 2003), Joseph Schwartz (1992), Hilary Putnam (1988 and 1992), Peter Medawar (1984), Robin Dunbar (1995), John R. Searle (1998), and Alan Irwin (1996).

Philosophers Martin Heidegger, Paul Sartre, and Jacques Derrida had worried their way through similar topics at a theoretical level in a complex, convoluted discourse of their own. They, too, weren't particularly concerned with the extended frames of reference that came with the extensions of the scalar spectrum. Where the thing itself is impossible to comprehend, the means of representation is unlike it - just as $E = mc^2$ is unlike actual exchanges of matter and energy. The equation gives only comparative amounts; it says nothing about how the conversions happen or what energy and matter are. Put another way, it is like Newton specifying the decrease in gravitational attraction over distance without saying what gravity is, and how it is that masses at a distance communicate.

The matching of words to things is only one of the difficulties the stretched scalar spectrum raises. Another is allowing the extremes to be more understandable, where the measurements can be known and tested only by those who have the appropriate tools. It is a long way from seeing the moons of Saturn and Jupiter up close through flyby cameras and understanding what they are and how they came to be. Given cross checks and testing, trust in objectivity isn't the same problem that Descartes raised concerning an undetected deceiving demon. Science denial, nonetheless, exploits whatever openings it can find to discredit what ardent belief denies. Karl Popper is correct to point out that a theory can't be proved right, only disproved. No matter how thorough the cross checking and testing are, that remains true. Mathematical theorems leave no room for doubt, but the sources and the rate of global warming do. Science goes by probability, not by certainty. However, the chances that theories such as evolution and the big bang never happened shrank considerably over the course of the 20th century. Those are typical examples in their exponential increase in probability.

Another traditional way to open up an improbable narrative is to say that what observation reveals has no bearing on spiritual matters. That is common enough to warrant a pause. The result is a bi-level discourse, which applies different standards of demonstration to physical and spiritual things. Descartes is one of the frequently cited examples of that body and soul duality, but Hume, Hobbes, and Locke likewise entertained a separated material world and detached Spirit. Nearly everyone still talks of body and soul as if they were two different things without saying whether the mind is one or the other, or what a soul is without a means of perception. Is it out of contact with things made of atoms, which includes everything in the visible universe? An internal division of body and soul is in keeping with projecting separate material and spiritual things at large, again avoiding accounting for their relationship. That has a simple remedy. On the individual level, merely rename the seat of consciousness 'the self' rather than 'the soul,' and keep it located where perceptions arrive and statements go forth - the place-into-and- from-which. When the feet move, it moves. Every integral object has just such a location.

Individual differences relate to that place somewhat, in the way numbers for x, y, and z change while the equation format, $x + y = z$, remains the same. The form is permanent, what it contains is subject to constant change.

Belief in a universal super spirit version of the spirit/matter bifurcation raises several other difficulties in the relation of spirit to material history. Long before the rise of 17th century empiricism, disorder and irregularity looked like anomalies in a universe thought to be infallibly created and governed. In Jewish, Christian, and Muslim doctrine, that was handled by assigning irregularity and disorder to angel and human transgressions. Even before telescopes and microscopes extended the perception of reality, the trouble with it was that it didn't account for a wrathful power punishing the good and bad alike, for the suffering and mortality of non-moral creatures, or for irregularities beyond the reach of people and angels. Nothing human or angelic is powerful enough to raise volcanoes and hurricanes, scorch deserts, or bring epidemics, for instance, and these beings having the power to do so sent philosophy back to square one, since any such super power then had to be as indiscriminate and unjust as it would have been had it created everything as it is.

The first wave of the rejuvenated empiricism made no attempt to work free of that moral and philosophic quagmire. As well as being brilliant in mathematical and material matters, Newton was notoriously prolific in theological ones. The first methodical and well demonstrated departure from the contradictions came, as I suggested, in the mineralogists Hutton and Humboldt, before Darwin made the harshness of nature in species evolution even more unavoidable. By then it was clear that confusion and imperfection in nature's proliferation were better explained by the naturalist terms that Hutton and Humboldt used and empiricism preferred. As physics has come to understand, the cohesive forces have too little range to make all arrangements orderly and geometric. Additionally, the electromagnetic and gravitational ones, which do have range, can't prevent eruptions, agitation from uneven heat moving from one body to another, collisions, and other kinds of perturbation.

The mechanics at both the micro and macro levels are now well known. Under thermodynamics, bonding forces also become breaking forces. Extreme heat separates parts of lighter atoms in convening selected parts in heavier ones, dispersing the excess as a byproduct, thus hydrogen into helium in the sun and generous amounts of light and heat sent flying. Gravity holds earth to its orbit, and at the same time pulls asteroids into contact with it - including the one commonly assumed to have killed off the dinosaurs and other species, one of several mass extinctions. By exercising an equal pull in all directions, gravity shapes molten masses into spheres, but as belts of debris illustrate it doesn't draw all celestial rubble into orbit around a star. Chunks of ice and other debris make up Saturn's rings, beautiful from a distance but from within a miscellany of disconnected pieces, nor does gravity prevent vents and turbulence in the interior and on the surface of stars and planets, cooling unevenly from outer to inner. Close up images of the sun reveal a similar turmoil, amplified, and breaking through the surface.

On the rock satellites, the outer result of uneven cooling is a wrinkled crust, and, on our particular planet, earthquakes, tsunamis, and volcanic heat plumes. Jagged mountain chains don't derive logically from a design or a master spirit, but from heat-powered natural causes under atomic matter invariables and matter/energy exchanges. The recently observed collision of two neutron stars, the leftover cores of disintegrated large ones, added to the visible kinds of celestial disturbance another case of an Einstein prediction based on constants, this one shooting off gamma rays and gravitational wave 'chirps.' Underfoot, and in evidence, are similar products of competing forces at work, open on the planet to air movements that bring rain, parch deserts, flood lowlands, and hurl typhoons at islands. Turbulence of that, and many other, kinds has never been a secret, merely something to be explained away by those who have faith in an altogether different cosmos, nowhere visible or demonstrable.

In Immanuel Kant, redefining the spirit realm as something called a mental noumenon offers a compromise of sorts between intentional design and the natural continuum. His *a priori* space and time originate in a nature that sets the restrictions. A master designer might or might not exist, but to Kant, who mentions such a figure several dozen times in *The Critique of Pure Reason,* it remains a way to account for nature's order but (as usual) not its disruptions. Explaining turbulence is partly what generated never-seen troops of spirits to begin with, from fairies at the foot of the garden to mastermind devils like Milton's, in the latter one kind of disruption falling to Beelzebub's specialty and another to Moloch's. That's not, however, Kant's way of handling what doesn't line up with pure reason. Like other traditional philosophers, he avoids confronting the problem of extended turbulence and wreckage.

By eliminating intangibles, naturalists like Hutton, Humboldt, and Darwin set a good many metaphysical, ontological, and epistemological enigmas aside. They too couldn't formulate a comprehensive theory of natural history. No one could see small nor far enough to identify what was needed to do so. They did, however, make major advances toward what became clearer in earlier 20th century astrophysics and geophysics, drawing on improved telescopes and microscopes. Makers of the telescope atop Mt Wilson deserve much of the credit for what Edwin Hubble saw from there in the first demonstration - that galaxies exist, soon followed by the realization that they were drawing apart.

Groupthink: Conviction With and Without Evidence

The habit of remaking the universe in human terms is difficult to overcome, but it can be done where groupthink is recognized for the regional phenomenon it is: a tool for coordinating group actions, building civilizations, and raising armies. No one has any great difficulty reading the fables of Homer and Virgil as interwoven history and fiction. Applying the same critical scrutiny to the cherished beliefs of one's own time and place is harder. That takes practice and dedication to empirical truth. In scalar terms, that begins at the particle level and extends through the buildup/breakdown levels all the way to the scattering galaxies. Seeing the continuum's scope and the power of its binding and breaking forces puts into perspective the extremely narrow range which lifeforms exist within. Asking *why* the continuum isn't a fruitful line of inquiry; asking *how* it works can be. In a universe which is largely inhospitable to lifeforms, conditions that sustain life would make a barely-visible, thin line on bar graphs that put sizes, velocities, and temperatures in proportion. To deal with global environment, we need to be more in sync with that extended history, and better able to resist quarreling and sometimes warring beliefs. It is probable that none of the species will be around for very long when placed in the context of the extended history of stars and black holes, but shortening the natural span of any of them out of miscalculations and ignorance would be unfortunate to say the least. Mechanized humanity has gotten a good start on that.

For the creature that has created the arts, performed acts of kindness, been mostly sensible in pragmatic affairs, and made improvements in medical imaging – not to mention sending cameras on flyby missions to the edge of the solar system – *not* to keep the planet habitable for its heirs is unthinkable. Yet, a sizeable number seem bent on doing just that. Fact denial holds high office, not just in the US but in many places. The best way to prevent it from going to disastrous lengths is through global efforts supported by national policies that put environment foremost. The reasons for doing that in all haste come built into knowing the full scope of natural history.

CHAPTER TWO

SEEING IS BELIEVING

The Argument

In the earlier 20th century, a consortium of sciences brought under scrutiny a portion of what had always been beyond human sight and hearing in establishing the micro and macro extremes of the natural continuum. As gaps widened between common assumptions and what could now be perceived, especially in sight and hearing, the scalar extremes created additional problems in comprehension and representation.

Time and Place

Despite agreeing with the each-in-every and every-in-each principle inherent in the natural continuum and the scalar spectrum, I like to keep three areas separate as much as possible: 1) the *what's out there* question, not extensively answerable until the past century, 2) *the receiving, recording and thinking equipment* we apply to it, 3) and the *means of representation.* I assumed these last chapter without making a point of doing so, but they warrant more direct attention now, as what are often collected under the categories of *objectivity, subjectivity,* and *rhetoric*. The second stage, the absorbing and interpreting process, is prone to making the self and the collective *us* overly prominent, as judged by the place humans and planet earth occupy in the visible universe. That is instinctive and genetic, since living things must look first to their survival and reproduction. It comes built into practically all levels of the biosphere.

To the third of that sequence, *representation*, a fourth could be added, a second reception stationed apart from it, as a viewer from a painting and an audience from a play. That presentations have to be intelligible adds conventions to the complex. These draw on common expressions or *memes*, a Richard Dawkins coinage for what comes packaged conventionally and is reissued countless times. Where genes dictate from within, memes coach from without. Because they are so readily acquired, they team up easily with groupthink. In extremes, this swings off into propaganda and brainwash. Where scalar extremes are involved, getting a given transmission on the right wave length for an intended reception isn't easy, and memes aren't of much help, where the measurements have never been in common circulation. Nothing in personal experience and not much culturally prepares for them.

The past several centuries, however, have cleared away a good deal of underbrush, exposed claims to magic, developed better ways to represent what isn't readily visible, and removed barriers to accepting what the newly visible evidence indicates. The result has been a wealth of charts, mathematical equations, and technical terms. Ethical and aesthetic philosophy have no need for these but natural philosophy does. Serious fact denial and evidence selection underlie inflammatory movements. Where science is common sense made methodical and extended, it becomes subject to critical review. The fact selection and denial of groupthink movements not only screen the evidence but often twist and misrepresent it. That is too obvious to need stating, but doesn't prevent the widespread practice of nationalism and blind faith from founding movements and electing the leaders of nations.

Brainwork

The extended reach of what was coming under scrutiny in the earlier 20th century was brain taxing. The mental equipment developed by human heritage in the evidence screening of the senses had, and still has, problems with much of what powerful detection equipment makes available. Getting a general understanding of comprehensive scales obviously becomes more difficult where the normal sensory range is left behind. What is known about how the brain gets from episodic sensations to words and concepts – fundamental to that problem – is in the keeping largely of specialists, but science writers like Gross and Harmon (2014), Arthur and Mitchell Bard (2002), Patricia Smith Churchland in neurophilosophy, Rita Carter in *Mapping the Mind* (1998), and the cognitive psychologists they consult can tell us a good deal about how we filter impressions, formulate them, and present them to others. I find *Cognition* preferable to Kant's *a priori* reasoning in that regard, because it is both a biological and conceptual notion. It matches something arriving with something already stored - in other words, usually *re*cognition based on similarity. It can do without language, but when something does acquire a name and is assigned to a category it becomes social. Cognition then becomes mutual, even general. Everyone who understands what the facts indicate is on the same page. When that process is fitted into a taxonomic scheme, such as the kingdoms and species one, it supplements the casual and the social with the conventional and the systematic.

The progress from sensations to brainwork, on to recording and eventually transmission seems simpler than it is, because it works silently and for the most part effectively within its customary range. In priority under analytic empiricism: first come instances and objects, and the sensations they raise; then classifications based on recognition and convention; and finally, on rare occasions, durable recording and explanations framed in axioms, equations, and theory. Western and middle eastern theology, reinforced by Plato, add a prior stage in making an Idea or divine Word come first, and only then does a material manifestation become available to the senses. In that sequence, reception 'ideally' sees through the materiality to its originating Form. Perturbations and disorder are sometimes then assigned to preexisting materials basically resistant to the ideal Form as the ultimate source. Nothing imperfect can come from that source. Thus, the entire material world, all hundred billion galaxies of it, can be set aside in the interests of the ideal Form, nowhere perceivable except through flawed objects. Any more realistic search for sources, precedents, and explanations, sooner or later, comes up against the insurmountable obstacle of an initial singularity - or whatever else one might call it. Needless to say, a supreme Source with an undefined purpose doesn't accord well with the continuum and the scalar spectrum. It would, logically, have to be assigned a source itself and that source another, or be left to mystery.

In chapter six, I'll go further into the sequence from sensations to facts and on to precepts in the reception, mental processing, and representation sequence in the context of Wittgenstein and others. All we need at present is to remember that under realism, as proposed by philosophers like William James, when someone sees a mouse the species name pops into mind as the scampering creature matches the classification everyone in the same language group uses. The classification could be wrong. The scampering creature could be a vole, but unless the sighting is a first ever or a confirmation of one by a zoologist, it will be of no particular consequence. In the context of evolution, objects and creatures lead backward through time to an ancestral source. That leads to another, and so on all the way to the first atoms and the stopping point, the big bang.

That's admittedly oversimplified. Observations range from simple citations of the 'mouse' kind to those of the 16th and 17th century, bearing on whether the sun orbits the earth or the earth the sun. That particular discovery involved prolonged scrutiny and accumulations of data that took several decades to work through. Within a more customary process of source finding and naming, recognition ranges from identifying the faces of friends to determining the causes of World War I. The latter is typical of what often occupies historiography, and in cases so tangled nothing very definitive is likely to emerge. Voluminous evidence and sometimes conspiracies are cited to explain what happened, focusing in World War I, for instance, on Venice, Sarajevo, Russia, and Germany, and in *Hidden History* (2012) on a cabal in England. Down among the details nothing very definitive emerges.

Pause on that level of complexity for a moment. It is complex as human events tend to be, but it does have underlying fundamentals that historians take for granted and needn't mention. Built into the firecracker fuse escalation of any large war was the nationalism that prepared for it ahead of time through indoctrination. We can see that everywhere, and have no difficulty linking it to the universal us/them polarity that runs through the animal kingdom. It often comes with an injection of paranoia, as seen in the run up to World War II in the brainwash campaign led by Joseph Goebbels, Otto Dietrich, Adolf Hitler which fueled a Marxist/Jewish conspiracy in order to exterminate Germans. Adding fear to collective fervor can cause a stampede, another ancient animal kingdom phenomenon. Animals that bunch up can also panic and take off. Fervent conviction in such cases depends on evidence selection and suppression, which is typical of nationalist indoctrination. It does the basic preparation.

What brought such a psychology about in *Homo sapiens* to begin with? That can't be answered with any certainty, but it is likely that herds of many kinds became us/them for somewhat the same reason. Placement in a time and place is universal: attachment and point of view come with location. Consider, too, that predation and the dangers it involves influenced genetic selection for millions of years. Birds flock and caribou travel in herds for a reason. Even an absurdly pompous figure like Mussolini can strike a match to the tinder that nationalism stacks up in any human version of flocking. We don't come to what specifically interests historians until well down the list of basic psychological and sociological factors. That doesn't usually include herd psychology. At the most fundamental level, our alarm systems are set genetically to go off when the right combinations come together.

What were cognition and recognition in *Homo sapiens* before language? Presumably similar to what they are for animals: identifying something familiar with not much more than a growl to express the recognition. Any of hundreds of kinds of creatures recognize the sun, but only the symbol user classifies it as a star and puts fictive animations behind it. Advocates of modular or automated brainwork sometimes expand the range of what identifies a cultural complex to include a portion of what we think we are choosing rationally. Among these are those Merlin Donald (2001) calls "massive modular hardliners". For them, a core self in charge of operations is largely an illusion. Our reactions are more automated than we realize. A physical action such as hitting a round hardball traveling 90 mph with a rounded piece of wood includes something unconscious or modular, something learned and practiced, and possibly something reasoned and analyzed. It comes more instinctively in some than in others, and thus can be considered more modular in them. That doesn't prevent improvement in them, too, with instruction and practice.

Brain neurologists tell us that the routes such perceptions as identifying a curve ball take through neural paths depend on their excitation value, and whether or not they call for an immediate response. Well-designed experiments can separate different kinds of episodes, and locate "concept cells" that formalize associations among sensations. They are high up on the abstraction ladder, where concepts are ready to be developed into theories. Language, numbers, graphs, charts, and pictures – common tools of systematic instruction – are owed to them, as is recognition that gets better with practice. Semantic categories like *kingdom* and *genus* in taxonomy function as molds in which data are firmed up, stored, and applied or shipped out. No one speaks of the animal and vegetable kingdoms without prior instruction. As for how the sequence from sensations to ideas works when it proceeds unhindered, sensory reception initiates neurotransmitter signals that pass back and forth from the thalamus to the basal ganglia and on to surrounding areas of the cortex.

As far as natural philosophy is concerned, nothing introduces mysteries or Ideas from transcendent sources into that process, nor do secret societies and cults have shortcuts known only to them. Forgetting – a vital part of the process – is mostly non volitional. We can tell ourselves to 'forget it' and, to some extent, put obsessions out of mind, but what remains and what escapes can be capricious. A catchy tune lodged in the brain can play over and over for hours while something we would sincerely like to remember vanishes. Rodrigo Quian Quiroga, Itzhak Fried, and Christof Koch find the process of selecting and discarding at the crossroads of consciousness and memory: "Concept cells link perception to memory; they give an abstract and sparse representation of semantic knowledge – the people, places, objects, all the meaningful concepts that make up our individual worlds. Their elegant coding scheme allows our minds to leave aside countless unimportant details" (*Scientific American*, *Remember When?* 2013). Shunning some things is a byproduct of forging links with others.

What consciousness and self-consciousness are in neurological terms and where they are located in the brain may never be settled to everyone's satisfaction, but the possibilities are clear enough to give neuroscientists

Koch and Susan Greenfield a framework to debate how neurons fire and are coordinated (*Scientific American*, *The Secrets of Consciousness*, 2013). However many may be modular and inherited, the brain has a great many tasks to perform, from silently operating the heart to thinking about Socrates. Brain scans track how some of the neuronal switches work and where the routing goes, and genetic studies can isolate genes that govern the development of the routes beginning in the embryo stage. Research emphasis seems to fall more on malfunctions and therapy than on such habitual matters as putting one concept in safe keeping while entertaining another contrary to it - a common way to be illogical and remain untroubled.

Gross errors in perception aren't rare, but they respond well to pedagogical treatment. The more habitual ones led Bacon to consider brainwork unreliable, where it hasn't been methodical enough in its procedure and been checked by peers. Because having the brain monitor itself affects the results, self-scrutiny isn't all that helpful in removing errors and in finding how we process sensations into concepts. Hence, just how ideas pop up remains an enigma even to someone in whom one has just now surfaced. That lack of access has led some visionaries and poets to regard themselves as prophets, especially in ancient times. No angels or demons were actually at hand, no archangel in a cave, no stick turning into a snake on a mountain. A good many such visionaries came forth in the Indus Valley, Mesopotamia, the Greek and Roman classics, Egypt, Persia, and Israel. Their persistence suggests that it is easier for fantasies to take root than it is to get rid of them once they have.

The brain's adaptability and capacity to focus give Gerald M. Edelman the title of his book, *Wider Than the Sky* (2004). The mind can cross the universe in a flash or flit about like Shakespeare's Ariel, touching down here and there. In Edelman's terms, the reciprocating ("reentry") process, in which one brain area refers instances to another, draws on technical language: "conscious processes arise from the enormous number of reentrant [reciprocal] interactions between value-category memory systems . . . and perceptual categorization. Through the complex shifting states of the dynamic core [the global brain], these interactions underlie the unitary property of conscious states" (77). *Unitary* derives from new exchanges being routed through accumulated experience. In developmental terms, "Only with the evolution of higher-order consciousness based on semantic capabilities do explicit concepts of self, of past, and of future emerge" (77). That degree of focal consciousness doesn't disagree with Daniel C. Dennett's notion in *Consciousness Explained* (1991) that a self "is not any old mathematical point, but an abstraction defined by the myriads of attributions and interpretations (including self-attributions and self-interpretations) that have composed the biography of the living body whose Center of Narrative Gravity it is" (426-427).

How brainwork developed to that level may never be known in detail, but apparently fingers and hands began drawing around the same time as the vocal apparatus began articulating sounds and brains began constructing grammatical sentences. Thus, crude cave paintings are dated a little before the 40,000-years-ago mark, an era sometimes assigned to the development of proto language into words and sentences. Whether in gestures, drawings, facial expressions, numbers, or words, the point would always have been to convey something that otherwise wouldn't cross the gap between minds.

Preconditioned Reception

We can concede that feeling and instinctive evaluation come with a point of view and still have some sympathy with Donald's impatience with a subsection of that topic sometimes called *qualia.* "Some snarls should just be abandoned. That's how it is, in my opinion, with the philosophical topic of qualia, a tormented snarl of increasingly convoluted and bizarre thought experiments, jargon, in-jokes, allusions to putative refutations, 'received' results that should be returned to sender, and a bounty of other sidetrackers and time-wasters" (369). No lack of evaluative judgment there! Attempts to convince people to accept evidence and act rationally run into desire, and into what others are thinking or what we think they are. Personal feelings, qualia, illogic, traditions, and various regional beliefs are formidable obstacles to the acceptance of what the evidence argues and rejection of what it argues against. Undistorted objectivity is hard to come by. We have to work at getting even halfway there.

That assumes that, by *qualia*, we mean not sensory data per se, but their reception and evaluation and not merely the "qualia-talk" that Hilary Putnam (1992) traces to Descartes. The reception of any sensation evaluates it. We can't get rid of that precondition, nor should we want to, but we can restrict the wishful thinking and partisan bias that often accompany it. Among the disciplines represented in a typical university curriculum, the sciences have the greatest need for uncluttered objectivity. That scientists themselves resort to rhetoric can obscure important differences in degrees of distortion. We've no reason to think that Hooke in *Micrographia* was covertly distorting anything merely because he had an ego and career ambitions. He was unintentionally distorting, however, in deriving flies from meticulous divine workmanship, ignoring most of the life of flies. That probably didn't seem a group matter to him, so entrenched and widespread was the assumption of a purposeful creation and so immune to evidence.

The real life of flies was easily filtered out. That isn't to underestimate rivalries and disputations among scientists or mistakes that stand for some time, rather to recognize that, sooner or later, the initial flurry over a hypothesis or theory dies down and additional data turn up supporting or refuting it. In fields subject to physical evidence, systematic development is usually the result - but obviously less so where the evidence remains scanty or, as with Hooke, where the general population believes something which the evidence denies.

As Jane Gregory and Steve Miller (1998) remark with respect to the public understanding of not only of science, but of most of what circulates broadly, "The media are an amorphous, interconnected, mutually dependent bunch. Word of mouth, cave paintings, hieroglyphics, tablets of stone, tapestries, manuscripts, town criers, books, newspapers, radio, cinema, television: all these have been in their own times and contexts, the medium for the masses; they have existed . . . alongside personal, private media that they have never entirely replaced" (104). That is one reason bad ideas go viral as easily as good ones. In that respect, Gustave le Bon's (1841-1931) once well regarded study of group behavior (1896) is in accord with Mark Antony playing upon the crowd in Shakespeare's *Julius Caesar*. Groups, le Bon maintains, are impulsive, mobile, irritable, credulous, and vulnerable to suggestibility. Yet without a doubt, he concedes, "human reason would not have availed to spur humanity along the path of civilization with the ardor and hardihood its illusions have done. These illusions, the offspring of those unconscious forces by which we are led, were doubtless necessary" (*The Crowd Study*, 42).

That big gods, too, arrive with big crowds is the conclusion of Ara Norenzayan (2013), but it seems likely that the size of what needs connecting rather than the size of a human cluster dictates the scope and power of the connecting force, whether real or fictional. Not too many alternatives are capable of accounting for the universe. Since no one knew of atomic matter invariables and the billions of years they had been at work, it was natural to put supreme powers to work making it, usually as it looked from earth with unassisted vision, while also usually subtracting what a given doctrine couldn't accommodate. The trouble with most such substitutions was, as usual, selective anthropomorphism. They were human centered, and the universe decidedly isn't. The hundreds of spirit forces that regional cultures have adopted have mostly been super humans, ignoring the disparity between them and distances in the billions of light years and masses millions of times that of the sun I suggested earlier. Nothing human works at all well throughout the scalar spectrum taken as a whole, from protons to supernovas.

I'll return to groupthink later, and be content here with one of the sharper and more bitter observations concerning throng beliefs from Ford Maddox Ford's character Christopher Tietjens in *Some Do Not . . .* (1924, 2013), one of the *Parade's End* series: Tietjens "fell to wondering why it was that humanity that was next to always agreeable in its units was, as a mass, a phenome- non so hideous. You look at a dozen men, each of them not by any means detestable and not uninteresting You formed them into a Government or a club, and at once, with oppressions, inaccuracies, gossip, backbiting, lying, corruption and vileness, you had the combination of wolf, tiger, weasel, and louse-covered ape that was human society" (70). In adding one more to that zoo, alarm spreads through people in clusters as it does in ducks, "contagiously. Like nations." Groups form into more than governments and clubs – ideological camps, religions, fraternal orders, secret societies, and revolutionary groups, to name a few. Ford's World War I tetralogy, written in retrospect about the hysteria that killed nearly 5 million and wounded nearly 13 million (1/5 of them civilians), could be applied to group fervor existing in nearly any time or place. No age, neither sex, and no large aggregate of people is immune to it. As le Bon insisted, though, the same ganging up also produces civilizations. It builds roads and raises towers. Though it didn't write books and paint paintings, its presence nonetheless encourages the individual efforts that have.

Equally feverish as Ford's contagious crowds, and sometimes as mad, are festivals and commemorative occasions. The crowds in Pamplona witnessing and participating in the running of the bulls are tightly packed and excited. The danger of getting trampled is part of the excitement, which, except for the active participation, doesn't differ greatly from fandom at athletic contests. This is one reason to call the influence of crowds *the stadium effect*. It influences collected numbers down to the size of those within the reach of voice harangue in public squares. The excitement of political rallies and rock concerts is usually momentary and controlled; it is a form of play. The fury of lynch mobs, raiders, pirates, invasions, patriotism, and righteous missions is another matter. With respect to how little evidence matters once an idea takes root, Daniel Kahneman (2011) observes that people can "maintain an unshakable faith in any proposition, however absurd, when they are sustained by a community of like-minded believers" (217). The greater the number behind a claim, the less its supporting evidence needs to be. That is doubly true when an animation subject to wrath is evoked and fear joins adoration in running through the throngs.

Rational agreement with a sound theory is quite different from that kind of swarm behavior, even though it sometimes looks much the same.

Consider the difference between something requiring demonstration and something based mainly on group fervor. The section of general relativity that proposed curved spacetime and gravitational lensing (the bending of light waves passing through a gravitational field) was received with enthusiasm because it involved predictions that could be tested and thus promised future research. It was also reasonably easy to represent as a field or fabric dented by a weight in the sagging middle, hence curving spacetime. The forward momentum of something in orbit makes it circle around the depression like a bike in a velodrome. That is an easily understood yet significant revision of Newton's action across a distance that required not only a force but a means of transmission.

In mathematics, a theorem is a step up in certainty from a theory, in being without exception or doubt and lasting as long as time even if no one knows about it. As Simon Singh (1977) puts it in reviewing proofs applied to Fermat's Enigma, a mathematical proposition is first argued step by Step, until at some point the evidence and the axioms leading to the theorem "are correct and the logic is flawless" (20-21). When that happens, the conclusion remains true forever. The scientific "poor cousin," as Singh calls a theory, could someday be refuted by counter evidence, replaced by something better, or modified. An adjacent area of study might also put a theory in a new light, as quantum physics revealed that classical (Newtonian) physics applied to objects at rest and in motion, but not to the building block particles out of which they are made.

Where science settles for high probability if it can get there, theorems in algebra, geometry, trigonometry, and calculus are unquestionable. The proportional relation of the sides to the hypotenuse of a right-angle triangle on a flat plane will never change. In applied mathematics, numbers become referential and incur the risk of counter evidence - a risk not necessarily to the validity of an equation itself but to its application. Place a right-angle triangle on a lumpy sphere, and the flat surface theorems for it meet with contingencies. The design of the trigonometric unit circle finds no exact match in roughhewn nature. It could be that, out of all the stars, not one is a perfect sphere. Indeed, with heat-roiled interiors and erupting surfaces, none can be. Convection currents and gamma ray emissions come with high heat. Hot plumes rise through even the layers of rock planets like earth. Different air pressures and temperatures bring air flow up to tempest intensity.

In terms of comparative reliability, where induction and theory fall below mathematical theorems, convictions based on revelations and personifications range from unlikely to impossi- ble. The history of the planet from a molten sphere to its current crunched up exterior raises no such problems in credibility. The moralized versions of natural events are often thought to be no harder to apply than finding the eye of a fly a marvel of craftsmanship. Several centuries of renewed and concentrated empiricism, however, have brought a higher regard for observation than was the case through most of the history of ideas. If the subject is flies, their entire profile is relevant. Spiders capture them in webs and eat them. Fly swatters squash them. They lead hazardous lives and die, well, like flies.

Applications

At least some of those who read this will have encountered one or more planetary watch- guards (that include the Worldwatch Institute, James Howard Kunstler, Richard Heinberg, Jared Diamond, Jeffrey D. Sachs, Fred Guterl, Karl Schroeder, Thom Hartmann, Paul Ehrlich, John Michael Greer, Lester R. Brown, George Monbiot, Bill McKibben, David Holmgren, Thomas Friedman, the Rocky Mountain Institute, the Union of Concerned Scientists, Carl Sagan, the League of Conservation Voters, Neil deGrasse Tyson, Elizabeth Kolbert, and others) aware of malnutrition, atmospheric chemistry, shortages of fresh water, and warming influence of greenhouse gases. Because of the wrap-around nature of earth's crust, its seas, and its air, environmental problems cross regional lines. Where they do, so must the responses to them. Commentary collected in *The Nature of Nature* (1994) and comparable publications cite what has become clear in climate reports and their long-range surveys: we aren't going fast enough to balance emissions of greenhouse gas with their capture and to tackle the 21st century crisis matrix in air, water, food, and population. The basic rules written in the chemistry and physics are well known. The potential level and timing of methane release is less than certain but is conceded by those who keep watch on thermometers and sea temperatures.

Can 195 sovereign powers proceed fast enough to avoid worsening the crisis matrix?

Probably not without a stepped-up educational campaign that contains natural history in general outline, and thus indicates the full extent of the scalar spectrum. As Lester Brown (2008) puts it, "Restoring the earth will take an enormous international effort, one far larger and more demanding than the Marshall Plan that helped rebuild war-torn Europe and Japan. Such an initiative must be undertaken at wartime speed lest environmental deterioration translate into economic decline and state failure, just as it did for earlier civilizations that violated nature's threshold and ignored its deadlines" (152-253). People will probably do what is nccessary eventually, but it took the bombing of Pearl Harbor to produce the World War II response of the US, and ice melts away quietly.

Greenhouse gases from the seas, termite mounds, swamps, stockyards, car exhausts, and coal- fired generators seep invisibly into the air. It comes down to how much risk we are willing to run when the penalties for making a bad choice are severe. Higher risks are tolerable when the potential damage is less. Where the damage gets as severe as rising seas and a smothering atmosphere, one chance in hundreds or even in thousands is sufficient.

No question exists that people in large and growing populations are putting nature under severe strain. Areas of prolonged drought, dying forests, and wildfires in western North America and Australia are strong, if not decisive, indications of global warming and regional drying. Trees die in excess each year west of the Rockies. Other indicators are water rationing in South Africa and more severe hurricanes in Florida, the Carolinas, and Texas. A Tim Barnett and Erin Brockovich-Ellis documentary, "Last Call at the Oasis," calculates that the better part of a billion people lack convenient potable water, and nearly as many lack proper nourishment. The yearly number affected by the storms, droughts, and fires is large and has been growing.

Urgent mitigation is clearly called for, and to meet the food, air, water, population crisis matrix we do have a "mitigation matrix," as Ernst von Weizsäcker calls it in *Factor Five* (2009). It proposes ways to improve efficiency in transportation, agriculture, waste disposal and reuse, building, water usage, and household management. The Rocky Mountain Institute has been at it since its founding in 1982. Workable programs, coupled with warnings, are the strong points of the Worldwatch Institute state-of-the-planet report of 2013, *Is Sustainability Possible?*, which finds the critical problem areas to be limited fresh water and food, a shortage of clean energy, and the emission of greenhouse gases. The authors warn particularly against methane sequestered in the seas that warming air and water will release. Even if we were able to rule out reaching a point of no return, which we can't, the cumulative effects of food and water shortages and of climate change were already severe enough several decades ago to call for urgency. Private projects are getting underway, among them the planting of a trillion trees. A good many governments and private organizations are in gear and moving. That the US took a retrograde turn in the executive branch in 2016 with cooperation in the legislature is undeniable, but the results are reversible. To move with the necessary urgency will, apparently, take more widespread awareness of the fact that the survival range for forms of life on the big stage is quite small and the chances of maintaining it shrinking.

CHAPTER THREE

YARDSTICKS AND LIGHT YEARS

The Argument

In one of its roles, natural philosophy brings together common experience and science in their respective and overlapping means of labeling and evaluating. In combining charts and graphs with mathematics and language, both it and science resort to the coarse grain measurements required by nature's prodigious production and measurement extremes.

< >

It is a false assertion that the sense of man is the measure of things. Human understanding is like a false mirror, which, receiving rays irregularly, distorts and discolors the nature of things by mingling its own nature with it. (Francis Bacon, *The New Organon*, XLI)

< >

Words and Things

The best and the worst thing early people did was develop symbols, especially syntactic and semantic language. I say the worst only because of how easily that prolonged development, so diverse, so powerful, can make the unreal seem real, can exaggerate, misrepresent, and vilify.

Wars of words usually precede wars of weapons. How did that come about? As historians and theorists of consciousness and language such as Merlin Donald (1991 and 2001), Steven Pinker (1994), Steven Mithen (2006), Derek Bickerton (1990), Terrence W. Deacon (1997), Daniel C. Dennett (1991), Ray Jackendoff (2002), and Harvey V. Sarles (1985) help us see, the anthropoid apprenticeship in the proto language of scratches, gestures, and sounds went a long time before language proper put words in arranged order. Judging by the number of languages and how quickly a segregated population changes one of them, different regions probably followed their own lines of language development. Stages of proto language on the way to grammar and semantics, and ultimately to cuneiform and written language, were, broadly speaking, a Bronze Age development - like the one assigned to Sumer's "proto-literate" period from about 3300-3000 BC. In other areas they continued to into iron age (1200-1000). The development of tools and weapons tells a parallel story.

Wherever and whenever words arranged in syntactic order came about, they enabled people to verbally cross oceans as easily as ponds, and to portray fantastic marvels as readily as real places and events. They journeyed vicariously into under worlds, pictured storm gods churning up the seas, and imagined human-like creatures flying in from far off places in the heavens. The Mesopotamian stories of the Eridu Genesis, the Enki and Ninhursag Sumerian stories of a lost paradise, the epic of Atra-khasis, the Enuma Elish epic of creation, Baal cycle, and Gilgamesh are filled with fantastic deeds and events. Some of the imaginary creatures and places were adopted for dedicated belief. It seems that most ancient cultures turned to that, probably in part because doing so added a sacrosanct level to ruling orders. Anything thought to be sacred commands awe, whether it is true or false.

Along with the pragmatic applications of language, and the invention of those and many other narratives, came a few attempts to scientifically and philosophically account for natural phenomena. The first of these on record took a philosophic, mathematic, and scientific turn in Turkey with Thales in the 6th century (BCE). Not a great deal that could be called methodical or scientific followed immediately, but over time it became evident that observation in effect reverses the forward arrow of time in that what currently flies, walks, or swims retrieves some species history. As we now know, in DNA and embryo development, what is present stretches even individual bodily timelines and brain development into the distant past. This isn't greatly different from the way visible galaxies stretch cosmology. Telescopic and microscopic scans, in going deep and far in size and space, also go backward in time. That is true in behavioral, as well as atomic, matters. In learning to walk, a child retreats to the race's original bipedal venture. In learning to talk, it reenacts stages from proto language to grammatical language: first babble, then words, then sentences, and thereafter, for a lifetime, continued additions to vocabulary. In these and other areas, the past remains present - if not as far past or as immediately present as light from galaxies that has taken billions of earth years to reach an observatory.

That doesn't make the retrieval or the understanding of scalar extremes a great deal easier.

Where what is to be represented comes in extreme sizes, distances, time, and luminosity, the gap widens between the object, customary perceptions of it, and representation. Can two dimensional holograms transcribe three – or in *M* theory, eleven – dimensions? Only with the understanding that, if the two dimensions are to represent more, they must be viewed differently than a flat surface parallelogram. How much of a species history shows in a current example? It can be much or little, depending on how many interim examples and how much biological or botanical history can be detected in it. Did the changes that produced the current bobcat come about as Darwin believed of evolution generally, through what Gould calls the "relentless accumulation of tiny changes through immense time" *(Structure*, 94)? Or in spurts with intervals of stability, as in the punctuated equilibrium of Gould and Niles Eldredge? Either way, the changes were incremental, came by means of inexact replication, and take us into distant history. As in a strobe light presentation, the appropriate sciences catch a glimpse, lose it, then catch another glimpse further along. The incessant change and gradations are part of what require lumpsum approximation.

As to the names that the sciences and common usage adopt amidst any long train of changes, when is a fish not a fish? 'Much of the time' seems the appropriate answer. The fossil record has several kinds of fish-ness, from creatures that swim in the ocean and have gills to amphibians that left the water and began exchanging fins for limbs (an extended process recently the subject of a televised program called "Your Inner Fish.") Fish with lungs *and* gills are among the in-betweens, as are land animals that returned to the sea or, like the spinosaurus dinosaur, were partly aquatic all along. Buildup and breakdown cycles are continuous operations. A dead and decaying fish or one being consumed has no conclusive moment when it ceases being a fish. Small differences make separating *Australopithecus* from *Homo* as difficult as separating land crawling tetrapods from amphibians. That would be even more so if the fossil record were better. Rather than one missing link, each discovery of an in-between adds two smaller gaps that might or might not ever be filled. One tendency of better data has been to extend the ages of some species, as, thanks to bones unearthed in Morocco, *Homo sapiens* in 2009-2017 added many tentative centuries to its portfolio.

With extended histories and minute changes, almost anything we look into has problems with the transitions that come with the successive nature of the continuum. Traditional problems, too, hang on. For followers of Parmenides and Zeno still haunting academic halls, time has the potential for infinite division in common with movement through space. By the third decade of the 20th century, moving land masses, quantum mechanics, and relativity had presented additional twists in establishing a comprehensive story, and with special and general relativity in the first and second decades, Einstein signaled further kinks in the relationship of science to the general public. Joining space to time and making gravity a product of spacetime curvature was one such kink - not hard to picture as a sagging web with a weight in the middle but not easy to get time, motion, and velocity hooked up in understandable terms. The speed of light in a vacuum has remained a constant, but the actual course light takes can be curved and impeded differently along the way.

Moreover, as physicists tell us, an electron in water can travel faster than light also in water, if not faster than light *not* in water, thereby presenting another rule with a footnote.

Add to these comparatively recent complications the traditional riddles of nullity, empty space, and void that have traditionally pestered philosophy. They remain as unsolvable as how something originally came from nothing - if it did. If it didn't – that is to say, if something was always there – we are once again caught an infinite regression. Recent, as well as traditional, philosophers have weighed in on that and on representation in general (among them Richard Rorty (1991), Robin Dunbar (1995), W. V. Quine (1992), Hilary Putnam (1988, 1999), W. H. Newton Smith (1981), Daston and Galison (2010)). The possible relation of something to nothing in modern cosmology has become twisted enough that John Barrow fills over three hundred information-packed pages of *The Book of Nothing* with it. For these and other reasons, as I pointed out earlier, accountability to unfamiliar things in the earlier 20th century became a leading language assessment concern. In brief, whereas traditional training in rhetoric concentrated on figures of speech, devices and persuasion, discourse theory now tends to swerve into ontology and metaphysics, where it encounters the anomalies of the continuum. Normally, the brain translates tangible objects within the optical range into words so automatically that we scarcely realize it is doing so. We see an object, think *maple tree*, and overlook the fact that nothing made of cells could perform that trick until probably about 2000 generations ago.

Other departures from the ordinary also demand attention. Time since the big bang amounts to a 70-year lifetime multiplied by nearly 200 million. It takes such numbing comparisons to get near total elapsed time. Multiplying earth's circumference by a similar amount carries almost nowhere into space, where light years in the billions each have to be multiplied by just under ten trillion to produce a kilometer figure. When astronomers consider how far the most distant galaxies might have scattered since they issued the light waves now arriving, they get a possible 93 billion light years from earth to the perimeter. Multiplying that by re (3.14159) would produce a circumference, if any such exists. In any case, estimates so extraordinary need to carry a generous plus/minus (±) factor. The distances and numbers could be reduced to a small fraction and still cause a household calculator to have a nervous breakdown.

Let us acknowledge at the outset, therefore, that practically all natural history lies beyond the unassisted human range. Since I need to refer frequently to its extended scalar spectrum in what follows, I'll abbreviate the far off, far back, and absurdly large and small as simply the vastness factor, or *V*. Microscopic sizes and numbers are as far outside human standards – for convenience call those collectively *H* – as supermassive black holes are beyond the mass of the sun. *H* means you and me, the subjects doing the observing and making the calculations from a given place. *V* means *objectivity* stretched to an extreme, most definitely not us. Even within the familiar range of *H*, nature's heterogeneity is so extraordinary that the relations of its parts can't be represented in detail. Disorder in patchwork disarray and heterogeneity – or prodigious bifurcation as chaos theory thinks of it – add to the complications. The crossing vectors of objects large and small are so numerous that indexing and cataloguing applied to them must be put in generic categories and the usual cascading or nested files. Martin Rees in *Just Six Numbers* adapts the symbol of the *ouraborus* (the serpent swallowing its tale) for the range from "the microworld of particles, nuclei and atoms" to the cosmos. As that ancient icon indicates, placing the world and its creatures in the context of something beyond numbering isn't a new idea, but adding quantum mechanics, oversize measurements, special and general relativity, and immense heterogeneity scatters what it symbolizes.

So long as we realize that boundaries can get blurred, catchall category names aren't misleading any more than the term *vegetation* is compromised by the many kinds of plants. Degrees of heat tell us something, even if the coarse grain numbers can't tick off every variant of molecular agitation. The scalar extremes, however, aren't so easily handled. The names and numbers can be valid and still not carry the conviction of commonly numbered and named things. As Gould remarks in the ambitious summing up of his career in *The Structure of Evolution*, "Among nature's vastly different realms of time, from the femtosecond of some atomic phenomena to the aeons of stellar and geological time, we really grasp, in a visceral sense, only a small span from the seconds of our incidents to the decades of our lives. We experience enormous difficulty in trying to bring these alien scales into the guts of our understanding" (674), especially when extended time contradicts or escapes revered beliefs. One saving feature in the labeling and filing endeavor is that, with some larger animals, we can identify individual variants and name one Mrs. Robinson and another Joe Dimaggio. Although the names don't reflect day- by-day changes, reputation and news reports do some of that. Photographs, anecdotes, and chronicles race to keep up, but never completely overcome the distance between things and how they are portrayed. The infants Miss Egbert (Mrs. Robinson's hereby assigned maiden name) and little Joe, if they were typical, bore little resemblance to the adults and saw their day-to-day stages vanish into thin air. A self here and now for anyone is relative, not only to a self sometime preceding, but in some versions of "simultaneous relativity" to all time treated as if it were summarily present, a variant of the each-in-every principle except in reverse, every-in-each. We, of course, don't apply extended spans to friends and acquaintances as we do to rock history and the stars, but the categorical distinction between past and future and between causes and effects holds true in common experience as much as it does in natural philosophy. What can't be reversed in time's forward movement isn't 'symmetrical' in the language of physics.

Ingredients stirred in a mixing bowl can't be unmixed. The dust flecks and sand stirred by a hard wind will never return to their positions, no matter how long we wait or how many more winds stir them. No one has any great difficulty understanding and accepting that. The buildup/breakdown cycles are based on it. New composites made of castoff pieces are never identical to former ones. What becomes the key to naming something has to be chosen from among competing features.

The trunk of an elephant and neck of a giraffe aren't what classify them; we rely on biology and zoology to spell out the evolutionary sequences and their DNA record.

The difference between a bizarre error and a brilliant deduction isn't always obvious. In the art of distinguishing and naming, quackery and genius are first cousins. Martin Gardner (1952) recites from the self-proclaimed genius Alfred Lawson, speaking of himself in the third person: "in comparison to Lawson's Law of Penetrability and Zig-Zag-and-Swirl movement, Newton's law of gravitation is but a primer lesson, and the lessons of Copernicus and Galileo are infinitesimal grains of knowledge" (69). The best safeguards against such claims are an acknowledged method and a consensus among peers. And remember, Lawson is right about the laws of movement as understood in the 17th century. They weren't good for squiggles and swirls. Put volcanic vents under a law of penetrability and acknowledge erratic motion and you find science being put to the test. These and other irregularities do exist, and so, at some level, natural philosophy must accommodate them. Double checking by peers isn't always enough to apply a classification to such things as penetrability and swirl movements.

As to Lawson's penetration of one thing by another and complex movements, they are what chaos theory takes on, knowing now as he didn't that they are the product of thermonuclear universals and exchanges of matter and energy. They follow other laws as well, such as lighter heated substances rising through colder heavier ones taking the course of least resistance. It is also true that Newton's measurement of gravitational strength is inadequate for several bodies in mutual attraction. Coastlines are often too irregular for designations in conventional east-west, north-south directions. Even so, turbulence and crookedness don't escape all measurements, as instruments demonstrate that gauge the velocity of wind gusts and air temperatures in the stir a storm generates. Science avoids some scrambles and handles others much as common sense does - by rising above instances in an ariel view. We settle comfortably into lumpsum approximations and have no reason not to.

Fortunately, evolution itself isn't susceptible to extreme turbulence except in eliminating species. It does, however, raise naming problems of its own, one of which is what to do with the residue of past species in current ones. DNA proportions now applied to taxonomy didn't figure in the pre-DNA assignment of species names. As the biologist and taxonomist Ernst Mayr remarks, genus designation and history are less certain than those of species (281), and the evolutionary sequence for a species or a genus is easier to establish than one for a still higher classification in the taxonomy hierarchy. Species are ancient, their genus more so, and so on through family, order, class, phylum, kingdom, and domain, each a notch further above instances. The higher designations are the animal, vegetable, and fungus kingdoms. In reverse of that, one small tip of the branch of nucleated cell life led to animals, a diminutive tip of which branched into mammals, a minuscule twig of which became *Homo sapiens* some 250,000 years ago. *Homo sapiens* is itself a highly selective term. In the context of evolution, a human is technically part tree as well as part fish. No species title acknowledges the full line of developments that every species has. That everything animal, vegetable, and mineral also has a stellar origin and falls under atomic matter invariables isn't directly relevant to evolution, but it is to nature's branching from simple to complex isotopes, and eventually causing some non-sentient matter to stir into life.

Uncertainty and complexity plague even common daily matters, such as weather reports. Air travels through the upper atmosphere in the jet stream, a 20th century discovery, cools in quiet downside pools, and warms when it pushes down canyons. It follows the lay of the land that has been undergoing incremental changes since the outset of the planet's air envelop. Ocean and air temperatures, topography, atmospheric chemistry and heat, and disturbances such as volcanoes also enter the calculations. Particulates in the air (SO_2) affect the weather and, over time, the climate. From the sum total of these and other factors calculated over centuries, climate may seem more determinable than next month's weather is predictable. That too, however, is doubtful given the number of long range factors, among them the influence of plate tectonic subduction, the tilt of the planet, slight eccentricity in earth's orbit, and unpredictable variants in the sun's radiation. Earth historians have to be chemists and allow for such improbable things as worms stirring the soil and releasing carbon, thus, as Michael E. Wysession suggests, helping prevent the globe from becoming a perpetual snowball. Human contributions include fuel use, the explosion of atom bombs, and particulates moved by winds. The latter can be surprising in circling the globe, as massive volcano eruptions and the impact of a large asteroid may have done in the dinosaur's extinction. Dust from China reaches not only the west coast of North America but the east coast as well. Winds from the African deserts carry nutrients to Brazilian jungles and thereby affect green growth. Winds from Mongolia carry iron-laden dust that feeds Pacific phytoplankton growth, good for CO_2 capture. In short, almost everything that moves through air or water affects the weather and thereby, over time, the climate. One car stirring the air driving down the highway, not much - a billion of them, however, quite a bit.

Epitomes, Zooming, and Foreshortening

In using one thing to represent many or in substituting one name for another, part-for-whole representation in metonymy and synecdoche strikes a compromise between remaining grounded in facts and flying over them at high altitude. Making an epitome of something is a form of zooming in and foreshortening that uses one thing to represent a category. This is also a useful device in leveraging familiar scales to get at extended ones. The surfaces of Saturn's moons reveal a history of both them and their planet. Since atomic matter follows universal laws, they also say something of earth and the rest of the solar system. In naturalist terms within our range, long-term processes are inscribed in undramatic rocks and trees, in redundant animals, in streambed cutting through rock, in ice-age scratches left on mountains. Any of these can be used to represent much more on and around the planet, and in speculation on similar rock planets. Under the every-in-each principle, nothing stands apart from the total. We mostly settle for types and categories.

In rocks as deposits of evidence, the record of past eras is disturbed, not arranged in the distinctive layers of the charts and diagrams but standing in stressed, folded, and broken shapes. The mountains themselves we can see directly but not the layers underneath. Since the strata are stacked chronologically, what an exposed layer reveals in one part of the planet characterizes other parts at the same level, and that presumably holds everywhere a formerly molten mass has cooled and hardened. Concerning specifically lifeforms, mountains tell geologists, paleontologists, and evolutionary biologists that amphibians and reptiles had their turn for 300 million years before dinosaurs and flying reptiles got in some 175 million. In measurements of that extent, lifeform evolution has been influenced by earth-orbit wobbles, climate, wormwork, plate tectonics, and volcanic carbon released over time. These and other factors have determined what has survived and what has gone extinct. In that context the bipedal predator that named itself wise is a rank newcomer on foot in the savanna, on mule or camelback, and eventually kicking up dust in motorized vehicles racing across the desert. Anthropology, history, literature, and religions arrived only a few thousand years ago, science and informed natural philosophy a few hundred. Still, more can be inferred from the record if not with certainty, as,, for instance that the physical weakness of humans caused them to team up and to extend their arms with clubs and spears. For whatever reason, humans became highly adaptable, inventive, and cooperative. Keeping such things in mind, if only as speculation, in some cases goes a modest way toward placing *H* in the timeframe and expanses of *V*.

Selecting specimens to be representative is the most common way to simplify rock and fossil history, especially in the Cambrian some 540 million years ago. In my living room, I have an approximately 360-million-year-old set of orthoceras fossils embedded in a marble matrix. To put that into perspective, that's over 1500 times the span usually granted for humans, and 5 times the span since dinosaurs. To the eye, the fossils look every bit as contemporary as the artifacts that keep them company - and, in a sense, they are - since foreshortening is one of the prominent aspects of detail selection. That class of fossils brings something with it of the era that spawned cephalopods in whorled chambers. These particular specimens arranged themselves artfully when they died, and thus became something like an arranged diorama mockup. Waiting to be quarried, polished, and displayed, they lay under terrain that lurched randomly before pausing under present-day Morocco. By the time I acquired them, craftsmen had placed them in an aesthetic, economic, and interpretive framework that includes jewelry, tourist trinkets, and other examples of lifeforms; ancient cephalopods meet modern marketing.

Eras often get superimposed one atop another like that. The advantages of foreshortening and going directly to an era 360 million years ago are obvious. The mind doesn't need the wormhole transport that science fiction provides for bodily time travel. The flexible brain, with its foreshortening capacity, classifications in tiers, and associative powers, can manage that quite well by skipping the in-between. That's what samples and tokens often do: transport us past intervening times and places to a particular segment of natural history lodged in rock in among others. Where the succession of integral bodies has, somewhere along the way, broken all the way down into gas and dust, it is harder to know what the preceding stages might have been. The trail vanishes in a cloud of particles, but the continuum tells us that the connecting line is always there. Planetary and stellar nebula mix various levels and ages of former compositions, interpenetrating, swirling, zig-zagging, hot and cold.

Changing the pace in slow motion and fast forwarding isn't technically a part-for-whole method, but in pauses it selects one moment to represent more. Movies and television advertisers have discovered that images can flash onto a screen at a rapid pace without confusing viewers and thus tell entire life stories in 15 seconds. Until they were outlawed, subliminal cuts registered somewhere in the brain without even proceeding through consciousness. The interception of two things taking different courses can also be revealing when we catch them crossing paths or colliding. In weaponry for instance, to choose an optical example, missiles can intercept a moving target at a combined closing speed of 15,000 MPH, and strike not just the designated target but a precise spot on it. Such an interception of flying missiles is comparable, a spokesman once remarked, to getting a hole-in-one with a golf ball struck in Los Angeles and finding a small, fast-moving opening in New York City. The universe has many collisions of high improbability in appearance, but mechanical and necessary under atomic matter invariables. Back on earth, socially speaking, under the influence of newsprint, cinema, and television, fads that once lasted generations now enter and leave the stage at an accelerated pace as the vernacular picks up and drops jargon in the flash of a split infinitive. The street-smart communicate with cocked heads and shortcut style, intelligences intercepting one another by means of a crossfire of symbols.

In science programs, zooming and retreating are common ways to adjust scales and charts.

They can get right to the point in showing, in a few seconds, the planet changing from a pale blue dot in the distance to a defined object seen from an altitude that shows the continents, and settling down in Times Square. Developing a mental agility of that kind, comparable to gymnastic flexibility, is one goal of practicing scales - based not on groupthink flash points and popular misconceptions, some of them fervent and equipped with weapons, but on comprehensive realism subject to displayed changes in scale. Getting a proper distance and simplifying the detail are vital to some shortcuts, generalizations and montage effects to others. So convenient and effective have such shortcuts become that more articles like Catelijne Coopman's "Visual Analytics as Artful Revelation" (2014) and entire monographs on zooming wouldn't be surprising, taking it as the opposite of stretching and using pauses as the opposite of accelerated skimming. Zooming in can be reversed, the point of view fixed distantly on something like the solar system as a whole. Back and forth manipulation like that is what histories commonly do as well; literary history, for instance, in spot-checking genres like the epistle and pastoral. A survey of them selects, then skips, selects and skips. If we want to take in the history of tragedy we close in on Sophocles and then again on *Macbeth*. Each pause is a closeup moment, a point of arrest.

Scale adjustment has always assumed that particular episodes in human history were part of connected longer stories, but the traditional overall stories differed from culture to culture and era to era, and were selective in evidence. Rather than explaining an actually existing universe, they characteristically proposed alternatives featuring humankind. Supernatural figures in them came in the shape of a Venus or gray-bearded Zeus, looking as if they were biological creatures in spite of their inventors and cultures saying that they weren't. Besides being immortal, they were often able to manipulate objects from a distance. Intrusions from beyond, bringing revelations, were once a common means of having an invisible or unknown realm intercept the known one, as when Hermes and Mercury drop into Greek and Roman affairs and angels into scripture. Instructions from abroad and stretched perspectives are narrative devices that Milton uses liberally in telling the story of a presumed universe from beginning to end. In *Paradise Lost,* divine oversight and omniscience enable an extreme foreshortening, in which divinity takes in all times and places at a glance - a confirmed foresight then, as if divine omniscience has been confirmed. What actually unfolds is the poet's narrative plan and a conventional alternative to natural history. With or without anthropomorphic superior beings, foreseeing and confirming are age-old narrative devices. The narrative function of messengers in that respect is to move the plot along and make connections. Everything unfolding down below is answerable to that animated each in every scheme, the inclusive 'every' being the divinity possessed of omniscience and omnipotence.

In that context, zooming and foreshortening alter how the perception of events affects us.

Tied to a mammalian body whose circadian rhythms developed over epochs, the human brain leaves not only details but emotions behind. Detail selection in zooming and fast forwarding has that effect; the plot of *Oedipus Rex* told in a few sentences doesn't raise pity and fear. In compression that is similar to televised weather forecasts that flash several hours of cloud formations and light changes across a screen in a few seconds, actual cycles of buildup and breakdown can be rapid or slow, but we know the regularly paced and sized stages only from experience. We can't take in the entire schedule of nucleon decay from fractions of a second to millions of years, but we are familiar with how the minerals of decayed trees go into grass on their way to herbivores, or into a robin by way of a worm. The telescoped or zoom form of all such transformations can be compressed into charts, as many are in the periodic table, nuclide chart, and a few other compressed examples, a couple of which I'll reproduce in a moment. Other spans are mind-stretching and, in combination, may be irregular and unpredictable. We have to approach them like a tailor keeping up with a runner to take his measurements.

Mapping the Confusion

Back to the world that is mostly in sight when assisted by magnification and detection equipment of great sensitivity and accuracy. Nature's heterogeneity and roughhewn configurations come at all scalar levels from the subatomic level to the dispersing galaxies. At the subatomic level, kinds of particles in the standard model are so numerous that even physicists as exceptional at explaining them as Lederman and Hill (2013) need pages of diagrams, and use nomenclature as odd as the left and right muon, spin and half spin, cyclotrons, synchrotons, linacs, **SCRF**s (**S**elf **C**onsistent **R**eaction **F**ield or alternatively sometimes **S**uper **C**olliding **R**adio **F**requency), D-zero, neutral re^{0}'s, red, yellow, and blue up and down quarks, and much more. The particle traffic would resemble a freeway scramble in an ice storm if we could see it - that's if the particles aren't few as some alternative physics suggests and their apparent variations owed to the frequency, amplitude, and speed of waves, or if they aren't strings with several dimensions wrapped up in them. The problem in representing them may be that they don't really resemble waves, fields, and strings. The pace of composites coming together and decaying can be adjusted in presentation, but each follows its own program. Any such course doesn't allow freely roaming quarks. Computer-generated bifurcation out to a frayed edge shows a limited degree of that process:

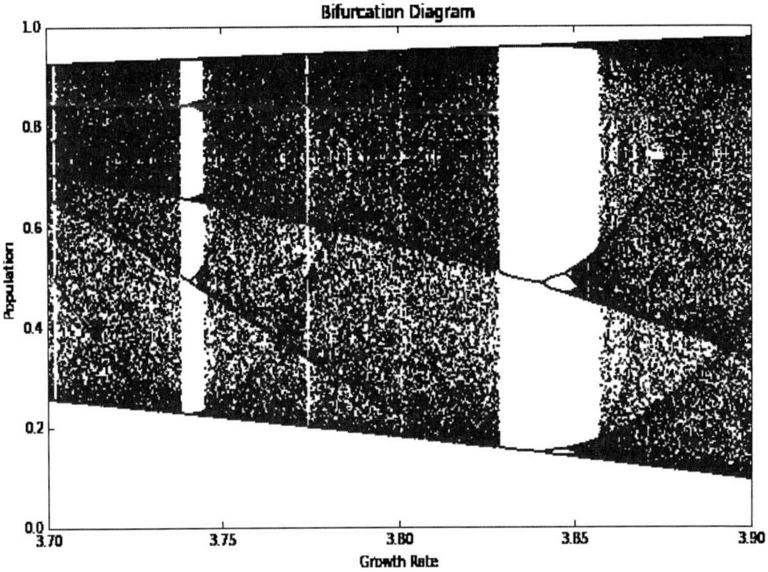

How complicated that can get at the building block level we can also see in the nuclide chart that puts thousands of alpha and beta decay calculations thus (from the National Nuclear Data Center):

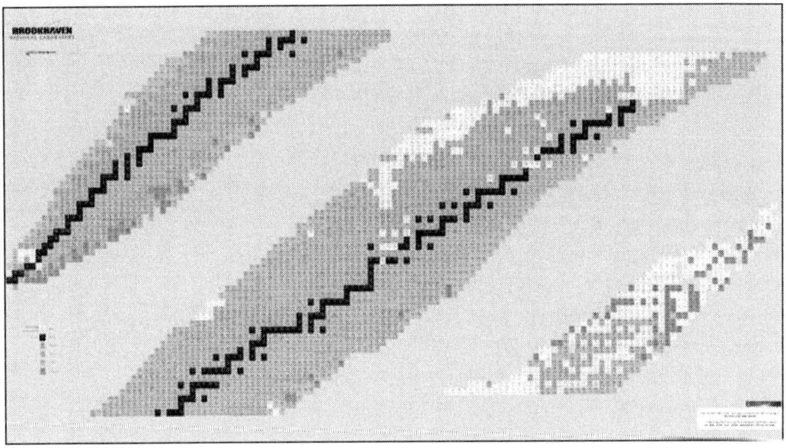

Every nucleon follows its own deterministic timetable. Phase changes at the molecular level follow a similar rigorous scheduling. The primary phases are gas, liquid, and solid, but the in-betweens include such things as mud, plasma, fog, dust clouds, and thick soup with lumps. The state in each case is dictated by molecular freedom and vigor of molecule movement. The charts and other representations are sometimes lumpsum approximations, much as class names like *canine* and *eucalyptus* cover dozens if not hundreds of variants. It is easy to see why a theory of partial chaos that can proceed further into irregularity comes in handy. The total disconnection of parts in utter chaos has no means of representation and no accountability, except for the overall areas that detritus and chaos occupy.

Astronomers can still be surprised by what turns up in limited confusion, as a crush of 14 galaxies did not long ago. The light from it took an estimated 12.4 billion years to reach observatories, so whatever pandemonium resulted will have mostly worked itself out by now (as, on a smaller scale, a good many galaxy mergers have done). In a colored version the convergence was a scramble in the making:

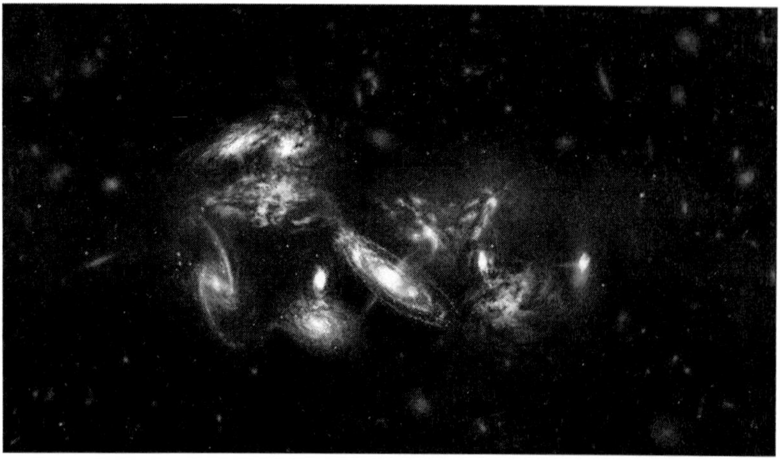

Rearranging that many stars and their satellites must have been one of the more sizeable undertakings of the universe - once the big bang and inflation had done their work and stars had converged into galaxies. The lesser mergers and collisions have come under the scrutiny of optical, infra red, and X-ray telescopes that filter out different repercussions. Up and down the scalar spectrum, the results have included a diversity of disturbed but not utterly disrupted patterns. Collisions from molecules to galaxies are another side of things that alternative notions of the universe either don't allow or use devilish trouble makers to explain. The symmetrically configured circle of Hoag's Object is of a kind that hasn't so far been adequately accounted for.

Its interior makeup is jumbled, but its overall shape is geometric:

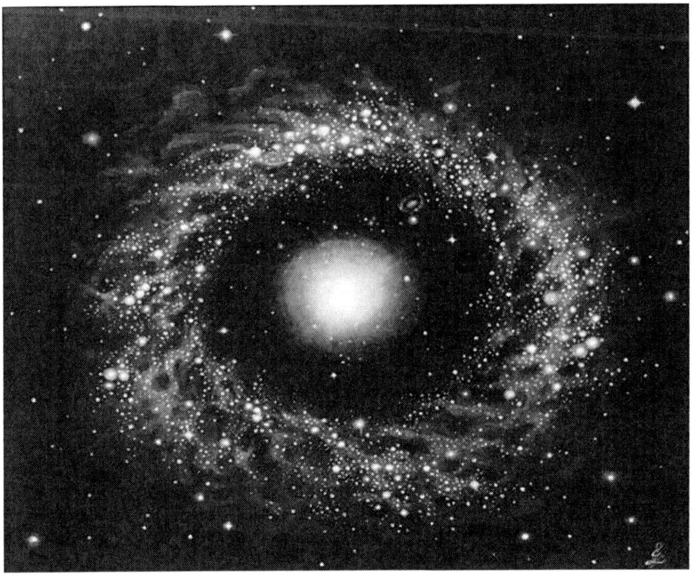

Another challenge to intelligibility comes in how gravity works *within* a mass. The complication is that, in addition to being drawn inward, the outer segments also pull outside bodies in as if from the core, thus, in effect, being two places at once. Another shortcut simplification is discarding most gravitational attractions in an extended field of them. Science took a long time to get from two body to three body calculations and will never gauge all such attractions at all distances. The movement of the core position is a more incidental matter, as earth draws the moon's core further toward it than the moon the earth's. Both are displaced by Jupiter, but there would be no purpose factoring that in - or any of the other displacements from around the solar system.

Growing with the Times

Since adjustments to *V* are personal in their particulars, yet in general outline more or less the same in everyone, in the next chapter I want to begin tracing the course of the symptomatic figure I mentioned earlier. Doing so is meant to sample the resistance that our instinctive subjectivity has to getting scattered here and there, as objectivity commands. Playing scales requires us to adjust to stretched and squeezed measurements, wholesale eventually under the every in each principle of the continuum, the located self being each and the natural continuum and its scalar spectrum every. We do that only bit by bit. Without a reasonably well demonstrated whole, we can't see how this and that fit into the whole. That's not the only perspective that counts, but it is one that natural philosophy, at some stage, has to reckon with. Concentrating on this one matter – learning to make scalar adjustments – isn't of course what characterizes traditional *bildungsroman* or growing up stories. Getting more accustomed to extended reality hasn't been a common maturation concept or goal of educational systems, and as a consequence many a misconception lives on, if not as prominently overall now as such substitutions for reality once did.

In environmental and political terms, McLaurin will eventually join those who come to see that the titles of books in the next few decades had best not echo those of the past few - warning of *immoderate* greatness, the *long emergency*, and *constant* battles. If they do, to echo still others, the *party's over,* that's *not the future we ordered,* and *dark times are ahead.* In that regard, I'll have him join the increasing number of those aware of what an expanding population is doing to the only known place that harbors life. Damage to the environment started being noticed and lamented in the 19th century in the romantic poets and in Humboldt, George Perkins Marsh, and Thoreau. Since around the 1970s, the chorus protesting that deterioration has grown as the evidence has mounted, especially what is happening to the planet's air envelope and climate. One important context for getting a perspective on that is how the planet, and especially mankind, fit into atomic matter dictates beginning the better part of 14 billion years ago.

CHAPTER FOUR

METHOD MEETS MISCELLANY

The Argument

In comparing familiar to extended measurements, scanning, sizing, and sorting bring comparative scales into play. Even for the preschooler McLaurin, seeing the opposite cultivated and wild sides of fences prepared him for a wide range of contrasts between the arranged and disarranged, and eventually for understanding universal buildup/breakdown cycles.

< >

The only material we have at our disposal for making a picture of the whole world is supplied by the various portions of that world of which we have already had experience. We can invent no new forms of conception, applicable to the whole exclusively, and not suggested originally by the parts. (William James, *A Pluralistic Universe*, 8)

< >

Matter rather than forms should be the object of our attention, its configurations and changes of configuration, and simple action, and law of action or motion, for forms are figments of the human mind, unless you will call those laws of action forms. (Bacon, *The New Organ,* LI)

< >

In the gap between the way the universe is and the way people conceive of the universe, there is much philosophy. (George Layoff, "Linguistics and Natural Logic," *Semantics of Natural Language*, 1972)

< >

Patterns Contrived and Natural

It is a nice second thought Bacon has in assigning fictive *forms* first only to the mind, and then to *laws of action*, thereby conceding built-in natural laws and recurrence independent of the perceiver. Methodical form and miscellany come mixed in both everyday life and science, in things human and natural. The extreme in method meeting miscellany is science as a methodical human product encountering turbulence as nature's chaotic extreme. Science and method have to turn away from anything that disorderly, science and turbulence being as incompatible as the tame and the wild.

Yet order and disorder are bound together; one demands the other. Every weed that grows is a system, following the patterns of its kind from sprouting to decay and disintegration - yet an acreage of weeds is a confusion. That same principle of intertwined opposites holds throughout the smallest of things to the profusion and confusion of stars and galaxies. These come in the hundreds of billions and lack an overall configuration. Thus, one component of almost any setting may be orderly in itself, but when joined to the others contributes to confusion. Though uncultivated nature is weedy, competitive, and often harsh, its individual recurrent forms give it enough identifiable patterns to keep the sciences and philosophy occupied and productive. Method meeting miscellany reached a new phase when early people began applying tools to roughhewn objects, creating order out of confusion. That isn't unique to humans, but comes as a new phase of systematic skill and understanding in them. Bifurcation graphs like the one reproduced last chapter represent natural recurrence without that intrusion, with continuing heavy lines through the multiplication and division that, in total make, nature chaotic. Civilizations testify to the order of human makeup, in constructing meaningful configurations out of nature's raw materials - buildings, for instance, out of quarried and chiseled rock. The rock followed a deterministic course billions of years in getting where it was found and put to use. Architecture drew upon that and made something intricately patterned out of it.

When we consider in how many ways and on how many levels we search for continuity in nature's proliferation, plus the patterns people devise, it may seem that not much is left of a day or a life that isn't devoted to doing just that. A child in encountering one thing after another learns the rules of the household, the classroom, the playground, and of nature. A doctor searches through symptoms to find the source of a medical problem. Historians card through documents to find the logic of successive events. Natural philosophy tracks through the entire natural continuum looking for recurrence in motion and novelty. In temporal terms, it is the locked-in continuity of natural history that enables the tracking of something both to an immediate source and to more distant ones.

Because sound procedures work the same for everyone and make use of common language and mathematics, methods for handling nature's profusion and formulating its enabling laws are teachable. That isn't adequate for natural philosophy's assessment of the continuum as a whole, however. Sunny Y. Anyang's *How Is Quantum Theory Possible* (1995) lists some of the typical difficulties that the building block level alone presents in quantum physics: "*The world exhibits many levels of scale and complexity that require radically different descriptions.* The classical and the quantum are two such levels, and there are many more beyond physics" (83). The 'beyond physics' side of that includes what, in university departments, falls to chemistry, in living things to botany and biology, and so on through paleontology and anthropology, the social sciences, humanities, and the arts. The evolutionary 'tree' puts the categories and subdivisions in terms of species. When the mainline trunk started branching, every new thing was only slightly different from its immediate predecessor. The count of slight variants at any given time amounts to a population census, since no two things are quite identical. We have comparatively little trouble with that or with what falls within our normal range in which a mouse is small and an elephant large. At the scalar extremes of the extraordinarily small and large, fast and slow, what takes one form or another requires instruments of detection well beyond normal human capacity to track down and formulate.

As to the scientific method often applied to nature's miscellany, methodical as it is, amendments and corrections are frequently needed as new evidence turns up. They can be as radical as a change of dating turning a presumed archaeological discovery into a paleontological one. After years of analysis the smashed leg bone of a mastodon unearthed in 1992 in southern California was proposed in 2017 to be evidence of a human, or maybe a Neanderthal. In any case, dating it at approximately 130,000 years old (a date 'fiercely disputed') would make it several times older than any hominid previously thought to have been there. That particular change may not stick - many don't - but if it does, it will match a recent unearthing of evidence of tribal life in eastern Africa that places symbol usage significantly earlier than it was previously thought to have taken root. That such revisions are only proposed and not established is characteristic of systematic study encountering suggestive but inconclusive evidence. It is also typical of any on-going interplay between circumstantial details and the methods applied to them. The overall goal for natural history remains the same, however: establish the succession of the continuum and locate when the more significant turns occurred. Where methodical classifications make defined breaks between adjacent phenomena, the things and events themselves often merge or overlap. How do investigative methods apply to data and assign them to theories and categories? After *hunch*, *speculation* is the least methodical of the proposal clan. We sort through impressions and data in search of patterns that can be pulled out and pasted onto chronological charts. Only when something stronger than speculation, something promising, turns up can it be called it *hypothetical*, a modest step toward a generally acknowledged fact.

Theoretical and *axiomatic* follow only where the evidence warrants. At the level of scientific theory, what may have looked miscellaneous or random at first turns out to be connected to something already established. With respect to the continuum as a whole, that means a logical place in the deterministic succession of atomic matter. To be clear about that, let us keep the overall sequence in mind. The earliest eras of atomic matter history became more visible in the ultra-deep field scans of modern telescopes. How composites were coming together in the initial star-gathering phase probably depended first on dust ball electrostatic attraction before proceeding to what is sometimes called pebble accretion. From there, gravitational attraction could start collecting larger masses, some of which are visible in light waves sent our way billions of years ago.

As to that phase, obviously a certain amount of scattered materials have to collapse inward to form the first and subsequent stars, planets, moons, and in-between masses sometimes labeled planetesimals or simply left unconnected. The following time and development chart has question marks at the outset (the bottom), because no one knows what if anything preceded the big bang. The Planck Era at 10^{-43}s is the briefest possible interval. It isn't visible, but atomic matter invariables underlie the theory that it existed. From there the accretion took effect at an extraordinary rate, thus:

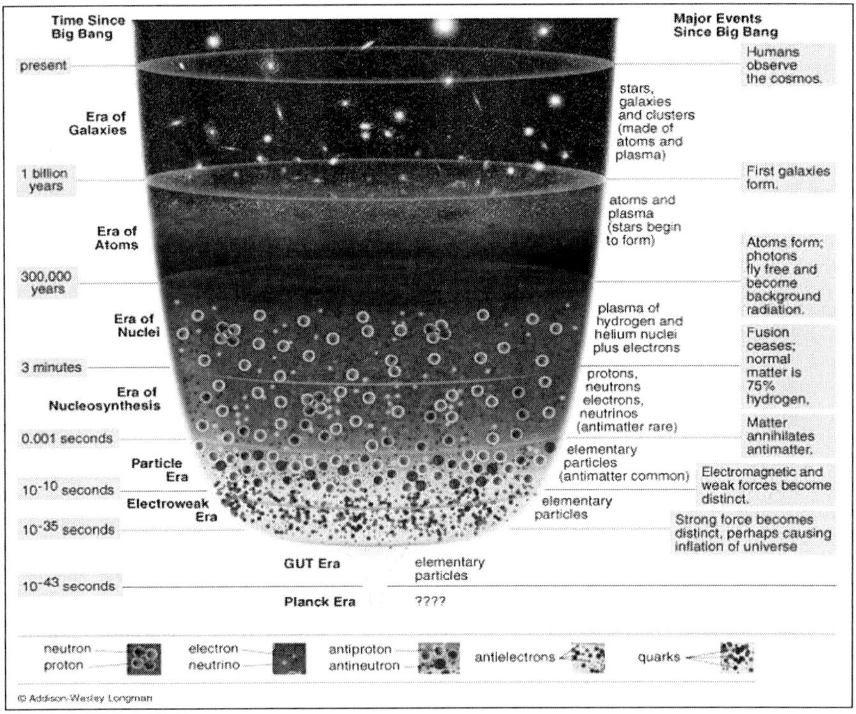

To set the stage for the relative placement of any given particular, we need some such chart of the whole with approximate numbers. The succession of composites built up and broken down is what carries through, and these are both deterministic and prolific in number. The disciplines that organize and classify them aren't as decisively separated as they are when sorted out and labeled in the lab and the classroom, because the stages aren't categorically decisive. In terms of disciplines, that means that physics overlaps chemistry, and archaeology phases into paleontology going backward and into anthropology and historiography going forward. Beyond the big bang outset, as we've seen, it encounters only the Nothing that some call Eternity, others an Abyss, and still others God. All these are beyond observation and knowing.

Induction and Realism

Confidence in accounting for what falls within the normal sensory range underlies empiricism and William James style realism, a variant of which Paul Feyerabend (1981) distinguishes from instrumentalism. Jamesian realism examines the object. Instrumentalism concentrates on how we handle it: "Realism and instrumentalism provide two alternative interpretations of science and of factual knowledge in general" (176). Realism assumes that knowledge of objects can be reliable, not, of course, that it always is. Instrumentalism pertains more to the descriptive phase of what's out there, and the reception sequence that follows presentations of it. For reasons similar to Einstein's critique of the Copenhagen group's instrumentalism, Feyerabend lobbies for realism (201) under which a change in a context or a means of testing doesn't change the object itself. That isn't to say that points of view don't count, merely that objects stand independent of whatever methods we apply to scanning, sizing, and sorting them.

The Einstein/Niels Bohr exchanges, which went on for years at the Solvay conferences and elsewhere, looked to be going mostly the Copenhagen way, with the discovery that observation at the quantum level changes the apparent nature of the particles (photons, electrons). In league with Boris Podolsky and Nathan Rosen, however, Einstein called the Bohr/Heisenberg uncertainty principle incomplete, and reaffirmed confidence that physical reality could, in fact, be understood and described - if not necessarily all the way down to particles. I won't go into the debate, the account of which in the *Stanford Encyclopedia of Philosophy* was updated in 2017, but it is safe to assume that the quantum zoo has unanswered questions, much as that article claims. In any case, quantum level uncertainty doesn't undermine Jamesian realism concerning composites above the particle level. When we walk around a sculpture, we don't question that it is the same object seen from different angles - we carry points of view around with us. Three people will see three different sculptures, but none of the three will deny that the sculpture is actually there and made of stone. Each onlooker has an invariable: the placement of the seeing and thinking, even as what each sees and thinks from that placement changes moment by moment. Testing and experience will determine whether the stone is alabaster or marble and may supply an interpretive context.

All the while, there the sculpture stands - rock solid evidence that what looks real is real.

Saying that much on behalf of Jamesian realism doesn't underestimate our perspectives, which are both individually and culturally conditioned. Nature's miscellany gets multiplied all the more by the number of angles that not only every observer but every social and cultural complex takes on it. Magnifying a particular aspect of something brings it into focus and puts the rest on hold. A given profile assigns to objects only the relations appropriate to the particular purpose, as an anthropologist puts *Homo sapiens* in a primate context for one purpose and a land animal context for another. *Context*, in that case, means frame of reference selected from the natural continuum. The placement of one object vis-a-vis others usually amounts to a definition or part of one. That two adjacent creatures, such as an elephant and a mouse, aren't closely related means that their common ancestry was a long time ago; a rhinoceros wading into a river has undergone many degrees of separation from the fish it encounters there. Repositioning it and the fish in the evolutionary line requires a time-and-change chart that goes back far enough to reveal their common ancestry. Juxtaposing an earth history chart with a history of matter chart gets us still further back, with time 0 the limit. The breakdown/buildup cycles, one of the near universals that carry through, should probably be more prominent than they are in natural history mechanics.

That they aren't may be due to science and philosophy concentrating more on what builds up than on what comes apart. Nonetheless, except for the particles initially and isotopes and elements that run throughout, building up and breaking down are equal in the making and unmaking of composites. That much hydrogen has never been broken down, and probably never will be, doesn't change the necessity of disassembling one thing to build another - disassembling four hydrogen atoms, for instance, to make one helium atom. At the familiar visible object level, we see the cycles in operation everywhere; in buildings that go up and deteriorate, in creatures that grow old and die, and in decay of a thousand kinds, the most basic of which is the level detailed on the nuclide chart. That not everything from a disintegration episode is picked up and reused shows in the solar system's asteroid belt, the Kuiper belt, and distant Oort cloud.

The largest level of breakdown/buildup is the nebular one, a subset of ongoing star and galaxy formation. This variant of the time chart depicts a level of natural history buildup prior to peak star formation:

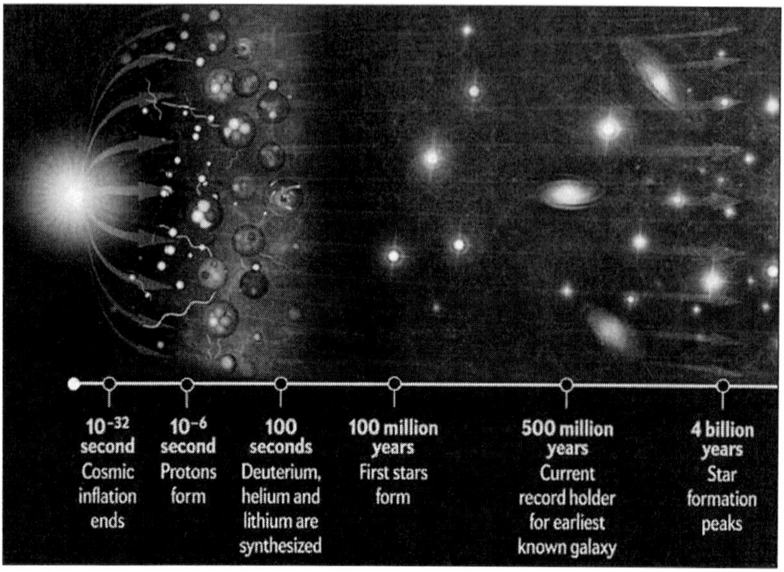

Natural philosophy is thereby reminded that, in whatever state it finds something, it has followed the succession. We might not ordinarily consider gas particles and specks of dust integral objects, but technically anything above the particle level is – atoms, isotopes, molecules, cells, chromosomes, dust, arms and legs. Further disassembling and reassembling are in store for anything currently composed beyond the basic elements. In some cases, the transition from one composite to another is known and certain. In others, nothing can be traced past the latest breakdown. Composites as they stand, before the next breakdown gets underway, are characterized by size, placement, and velocity. If velocity seems more important on that short list than is warranted, it is because movement distinguishes one integral object from another. To show that in special relativity terms, Einstein underscored comparative velocities. In common sense, that means simply that a forearm and hand pitching a baseball move relative to a planted foot and a propelling, slingshot upper arm. The launched baseball is then distinguishable from everything around it not only by its material makeup but by its velocity and direction. Though quite loose in its outer components, the solar system moves as a unit within the Milky Way. That galaxy in turn moves as a loose aggregate with respect to other galaxies. Every object on the planet shares the movement of the planet and, at the same time, moves relative to other integral objects on it.

What holds together affects some of the parts that have gone into the mix, as a cake batter in becoming a single bowlful of ingredients alters some of them. Patterns figure in miscellanies as well as in tight groups. That may seem surprising, but the adjacent parts of a weed patch do sometimes influence one another and the patch has boundaries. What characterizes rubble is the detachment of pieces that don't move in unison. As the components of an integral unit increase in volume and number, the encompassing order grows less tidy. Galaxy clusters are less geometric than spiral and elliptical galaxies, which in turn, with the exception of rings like Hoag's Object, are less geometric than spherical stars and their satellites. In order to look integral, composites have to be seen from a distance - a considerable distance when it comes to galaxies. No viewing distance shows the universe as a whole, and neither tradition nor common usage sanctions titles for it like the Great Welter, though that seems to be what it is; it has integral clusters inside other clusters, composites within composites all the way down to quarks, but no discernible overall relation of its units. The *uni* part of its name refers not to its constructive integrity, but simply to its being the only known conglomeration of galaxies derived from the same singularity.

Because mass and movement include such things as riffles, and qualities such as bristly and smooth, descriptive terms often accompany measurements of size and velocity. We feel more comfortable with the object when we can describe them geometrically as cones, pillars, spheres, semi circles, rectangles, flat planes, tetrahedrons, or squares, such standardized shapes indicating law and order. We are also acquainted with many rough forms and broken configurations. How do lumpsum estimates work when we set them beside such shapeless phenomena? Choosing what is pertinent and representative is the key to that. Take something familiar like a waterfall. It drops a certain volume of water a certain distance, but gradations from plunging drops to mist blown by the wind defy calculation. The falling water is simultaneously coming apart as a stream and composing a single waterfall. We've no need to know the moisture content of the air coming in and going out, or gradations of temperature on the fringes on a hot day and a cold day. Since such details aren't important, sizing and sorting can ignore them. The sheer scatter of a breakdown makes the pieces trivial, which is another reason for methodical study to ignore the breakdown part of the cycles, even though the mechanics of the continuum wouldn't work without it. Members of a clan or tribe that have scattered aren't of much interest to sociology beyond the reasons for their scattering.

Becoming too meticulous has been one of the commonsense complaints against science since the 17th century empiricism resurgence. In their fascination with microscopes, some of the earlier proponents of the inductive method blurred the line between significant detail and trivia, what needs to be gauged and what doesn't, the justifiable part of Swift's mockery in the 'naturalist' version of fleas-on-fleas ad infinitum. Relevance has to make the first cut if we aren't to fly loose. In everyday experience, curiosity and attraction add other matters unrelated to importance, as when our eyes track moving objects for no apparent reason. In terms of survival, movement probably signaled friendly or unfriendly presences often enough to give such tracking a survival advantage. Idle curiosity, in other words, isn't always idle. We do know, or can surmise that, where predators roam and landslides come down mountains, alertness and keen senses have an advantage. Thousands of asteroids can orbit the sun and remain a matter of indifference to everyone but astronomers. The large one headed our way is the one that counts.

The difference between methodical investigation and casual curiosity is usually the presence of a benefit or a threat. Creatures from ants to elephants have reasons for what they do; at the primitive animal kingdom level, that is nearly always the avoidance of danger, reproduction, or the procuring of food and drink - the driving forces of creature evolution. In our scanning of the natural continuum, the sample that illustrates a principle is the one that counts most. More often than not, the principle will be the buildup of a composite. What the object is while intact, meanwhile, shows in its movement at a velocity along a vector. All manner of scanning and sorting, X-ray, infrared, optical, and radar, concentrate on integral objects.

Moralizing and Humanizing

Anthro: human. *Morphism* form or shape. *Anthropomorphism*: attributing human form and feeling to a deity or something impersonal.

Just as troubling as miscellany for proper scanning and sorting are gross errors, one of the prime sources of which we've seen is the anthropomorphic instinct. Our fictions populate the universe with animate illusions. Every time we address an animal as if it understood more language than *sit* and *stay*, we are fictionalizing it to a degree. Indeed, we humanize nearly everything from pet birds in cages to the sun, moon, and planets named Mercury and Mars. One example familiar in the Middle East and West will do for illustration, but innumerable ones exist. That these differ by region, and sometimes take to battlefields, is enough by itself to question their correspondence to anything real, but their incompatibility with the natural continuum as a whole is the main mark against them. Local tribal legends we have no trouble setting aside as conventions, but incredible stories that remain influential in one's own culture are another matter. Anywhere from two thirds to four fifths of people in most regions cling to age-old fables contradicted by the age, expanse, and mechanics of the universe, and by what is obvious around the crumpled perimeter of the planet.

The only literary text in English that tells a complete story from beginning to end, cosmic and predominantly human at the same time, is *Paradise Lost*, and it encounters an impossible problem in matching its biblical and classical sources to what is plainly there. The philosophic prominence of St. Augustine, Thomas of Aquinas, and other commentaries of like mind were based on that same account of human origins, the history of the world, and its fiery end. In accounting for why so much of the creation has come to confusion and made living things mortal, Milton also borrows from an ancient Prince of Darkness tradition. Putting the rough and tumble side of wild natural history into such a framework was always questionable, since it required an unjust, wrathful deity killing off plants and animals wholesale and, even worse than mortality, afflicting them with pain and disease. In Milton's version of the story, turbulence in every category derives from a failure first in the angel ranks and then in humans to observe the one matter that truly counts, God's will, the source of everything.

One reason for the prominence of such a strong reaction to disobedience was probably the great chain of being, another people-oriented myth that placed living kinds under human dominion. Seeking knowledge beyond an assigned level upset the order, which, in class terms, was stiffly hierarchical. Eve in particular wants to know more than she is allowed and thus become more, meaning mainly equal or superior to her husband. In Milton's version, Satan rebels because he is offended by what he considers a personal affront in the promotion of the Son as Word and Light, which seems to him to lower angel ranks and him in particular. In a soliloquy, he later concedes that he was wrong, but he doesn't speculate as to how that flaw in his character and in a third of heaven's angels got there. Milton lets that pass, but the answer from outside the poem is that moralizing and humanizing natural history allows no choice. Something had to have gone seriously wrong if a universe thus infallibly designed and ruled contains so much in disarray.

Until the rebellion in heaven, revelations from on high make the congregation of angels ecstatic and coordinated in intricate dance movements. In the rebellion, a canon throws them into angel-flock rubble. That is the first stage of the world's ruin, an upside down or inside out version of divine revelations. Satan and his cohorts can't make or invent anything entirely original, but they can parody and upend what exists. Thus, the mouth of the canon comes forth from concealment in the reverse of an announcement from the Word: "By thousands, Angel on Arch- angel roll'd" (6.594). To that consternation Satan adds a derisive public declaration, another verbal opposite of the gracious Word:

> O Friends, why come not on these Victors proud? Erewhile they fierce were coming, and when wee, To entertain them fair with open Front
> And Breast, (what could we more?) Propounded terms Of composition, straight they chang'd thir minds, Flew off, and into strange vagaries fell,
> As they would dance, yet for a dance they seem'd Somewhat extravagant and wild, perhaps
> For joy of offer'd peace. (6.609-617)

That this scattering foreshadows the battlefields of the epic tradition makes *Paradise Lost*, in an odd way, the first of its kind despite coming centuries after the Mesopotamian and Egyptian myths, *Gilgamesh*, Hesiod, the *Iliad,* the *Odyssey*, and the *Aeneid.* The belated poet has, in effect, circled around behind his predecessors to tell the cosmic story. Misfortune would be entirely Satan's doing, except that he doesn't have the power on his own to raise mountains and freeze the poles.

To get enough damage, Milton must settle for rampant injustice and have God become indiscriminately destructive. There seems to be no other way to escalate a human affair into a cosmic one, which is what anthropomorphism calls for. What people do must somehow raise mountains, carve out canyons, and cause tigers to devour lambs.

Other problems that accompany the humanizing of the universe don't need attention here, but Milton does manage to go a step further than most who struggled with the impossible task of making nature's harshness and disruption look logical. The question as to why so many angels would have been so deeply flawed, he answers in *De Doctrina Christiana* by having free will open the way. Creature intelligence can't be strong and complete unless it makes moral choices.

Sufficient to have stood but free to fall is how *Paradise Lost* (3.99) puts it. Making free will the source forces the reluctant reader to live with the consequences. That, too, however, isn't foolproof, in that free will could have been exercised in ways that didn't involve disobedience but did involve moral decisions. That will apparently be the case once order is restored and the damned are separated from the blessed - or paradise restored will be vulnerable to further lapses for the same reason the initial paradise was. Thus, the dilemma: if moral choices have to be restricted eventually, why not initially? If they aren't restricted, what happened once can happen again and again.

That is enough to suggest the quagmire philosophy plunges into in squeezing the universe down to human size, even with the amplifying help of human-like gods and devils. In Sumerian, Babylonian, and Egyptian traditions and in Greek and Roman classics, polytheism had several gods and goddesses whose crossfire served a similar purpose. As Jeremy Black and Anthony Green demonstrate in *Gods, Demons and Symbols of Ancient Mesopotamia* (1992), the Mesopota- mians came up with different imaginary chapters of world history. They began doing so, as did Egyptians and people in the delta of the Indus, many centuries before the Hebrews, who in turn were nearly a millennium ahead of Christians and Muslims. None of these storybook universes explain nature, but they do distinguish one nation and one belief group from another. By inflating the repercussions of human conduct, their variants give their nationalist governing systems leverage in making what offends the civil order also offend the cosmic superstructure.

Such fables continue to characterize cultural differences around the globe, making sacrosanct what is actually an expedient for coordinating populations.

Mark Twain gives another twist to the Hebrew/Christian/Muslim world story in tracking Captain Stormfield through space to the conventional restored and improved paradise. The problem as Twain poses it lies in the logistics once death and disintegration are removed; after traveling billions of miles in a homemade space contraption, Stormfield finds heaven to be a scene of confusion, not from demons firing canons but from the overwhelming number of the saved and their standard issue paraphernalia. As one might expect with time endless and nothing to do, those who have grown tired of wings, harps, and halos are discarding them and milling around aimlessly. Other oddities in traditional paradises may be avoidable or deniable, but the numbers aren't. Either the saved have to collect in one place and be there forever, or treasured former attachments have to be severed and the billions broken down into manageable enclaves. One conceivable alternative would have been a heavenly regionalism, separating former allies and loved ones but allowing sufficient travel so that no severed connection becomes permanent. How the exchanges of the billions would work without confusion leaves the horde about how Stormfield finds it - except with more traffic.

Putting lambs in the prowling territory of predators does something equally unavoidable in challenging conventional doctrine. William Blake points that out in asking of the tiger, "What immortal hand or eye/ Could frame thy fearful symmetry" and following with "Did he smile his work to see?/ Did he who made the Lamb make thee?" That is a mystifying triad of questions for anyone bent on humanizing a universe which includes both tigers and lambs. Such a smile on one pleased to offer lambs to the slaughter wouldn't be pleasant to behold. *Paradise Lost* tries to work around that particular problem, among several that beset the system, by making animals with fangs and claws playful browsers and grazers in Eden. That their rending and tearing equipment is ill suited to a diet of fruit and vegetables adds just one more factor to what has to be passed over. Having the claws and fangs prepared for what lies ahead after Eden would underscore the foreknowledge of what is about to happen and letting it proceed. Natural history continues to go one way, notional contrivances several other ways, none of them accounting for more than a small fraction of what exists.

Curiosity Sets Forth

We've gotten far enough into the scalar spectrum to see what some of the problems of adjusting to it are, and have set aside one typical moralizing of a universe far too vast, rough, and aged to assign humans a prominent place in it. The scope of the problem is clear in any of the diagrams and charts of the natural continuum. Where do humans fit in the glass-shaped time chart pictured earlier, running from 10^{-43}s to now? This may be an appropriate time to introduce the template example of the scalar adjustments everyone has to make a little at a time. The overall course of someone stretching in concepts and measurements is progress in objectivity.

Discovering what's out there begins early and never ceases. The typical course goes from the instinctive subjectivity of the child to an experienced interplay of self and other, guided in this case by a growing appreciation of what science and natural philosophy say about the continuum.

When McLaurin, as a young roamer, is outdoors in the hills or along a river that runs through the family acreage, much of what he encounters seems a confusing miscellany. That isn't just because he is inexperienced, but because that's how nature in the rough strikes everyone. Confronting it has been a prolonged human experience, as it would have been also for pre human hominids for going on 5 million years since their separation from the simian line. If McLaurin eventually gets further into the natural continuum than anyone could prior to modern times, it is because he benefits from what improved instruments have been showing trained specialists since the invention of magnification. Thousands have been at work filling the glass of "time since the big bang" and its "major events." I'll have two lines converge in McLaurin's later development, his course and the development of a Theory of Everything (TOE). At the micro end, the findings of several decades collect into the standard model's subatomic level - but he won't encounter that until he is well into a post graduate career. He will need some mathematics to get his impressions properly sized and placed, but trigonometry and calculus won't be necessary for that - algebra and geometry will do.

In childhood, McLaurin encounters mostly what everyone does, though a rural setting in the Willamette Valley gets him out and around on the fringes of the wild more than most. His eventual exposure to the valleys, forests, mountains, deserts, and ocean of the west coast from Alaska to southern Mexico gives him a good sample of the planet's surface and nature's wilderness productivity. Areas of rough miscellany lie just through the fences that separate the wild from orchards, gardens, and fields. Coastal mountains of oak and fir overlook the west side of the valley. On the east side, the Cascades, stretching from Canada to northern California, are dense with fir, cedar, hemlock, and undergrowth on their rainy side. Forests of yellow pine are more open on the inland side, where aspen and ponderosas lead down to junipers and on out to deserts. Plunging streams cut through steep canyons on both sides. A variety of undergrowth competes for sunlight and root space. Nature overall, especially along the coast, is beautiful, has the usual ferocious predation and an abundance of growth and decay.

In the Oregon sections of the coastal and Cascade mountain terrain, the watershed on the west side of the Cascades collects in the Rogue and Willamette rivers, and on the east side in the Deschutes headed north and the Klamath headed south and west. That drainage system accommodates cycles of rain and runoff more bountiful than average but otherwise typical of how the planet's gravity-powered watering systems work. The uncultivated areas haven't changed greatly since people first migrated out of Africa and scattered in clans and tribes around the globe, hunting and gathering as they moved. The geological ravages the western continent has undergone show plainly and are by no means over, despite the gradual cooling of the planet and decline of volcanic and seismic activity. Reminders of deep time are all around.

The global transition from foraging to farming and industry was delayed in the Americas by a number of factors, remoteness from Asia and Europe chief among them. Except for the Rogue River area, the west side of the Cascades made the transition with less turbulence than the east side, where disputes among the Paiutes, Columbia basin tribes, and Nez Perce - and between these and incoming shepherds, miners, cattlemen, and farmers - lasted a good many centuries. Awareness of the last phase, between Indians and Palefaces, was still in the air in McLaurin's youth and was part of his introduction to wilder times. His grandparents talked about it as something their parents and grandparents had witnessed personally. Within some three generations, indigenous tribes, inheritors of about 240,000 years of the foraging way of life, had exchanged tent and longhouse culture for woodframe houses on reservations. Their economy in the past half century underwent a complete overhaul that featured gambling casinos serving lunches and dinners not of bison, fish, and venison but of beef, pork, and chicken.

Some of the planet's typical geological disturbances and of that economic make-over came compressed in an area that could be crossed by car in a couple of hours. In retrospect, after a few years of experience in scale playing, McLaurin will reconsider relics left by the economic makeover among the local Kalapuyas, Yamhill branch. These, too, wedged openings into the deeper past, in that case in anthropology that in turn led into the still more distant eras of paleontology and, ultimately, the geophysics branch of atomic matter. The latter was on spectacular display not far off, in snow-clad peaks and volcanic cones. What had been childhood curiosities will eventually become trail markers. After selected university courses even his former rambles through the fields, pastures, and woods, with their rare sighting of wild turkeys, coyotes, bobcats, puma, deer, and elk, will become a memory trove of illustrations and way markers.

Begin more exclusively within the sensory range that began stocking that collection: one of McLaurin's earliest recollections was of the primitive look of cows. Even at the time they had struck him as odd. Despite the kind having been shaped selectively into productive marvels of milk, hide, and beef on the hoof, he saw hints of wildness in their furry covering and the upturned horns of the heads. That impression had come from looking down on a stanchioned row of them from a hayloft. His more usual view came at the three-to-four-foot level of crooked hind legs and twitching tails. From there, the placid bovine eyes tilted outward the better to see grass for their broad teeth to crop. He wondered if the brain in the thick skull, inherited from aurochs, might somehow still have phantom memories of fang-toothed smiledons and clawed agriotherium. Probably not because, unlike antelope and caribou, cows were no longer wary. It was true that they *could* stampede, and he saw it happen once, but they panicked less readily than sheep and horses, and were more like the truly odd looking moose in that respect. Even on that one occasion, they didn't go far before turning about and staring at the passing band of coyotes that had startled them. As he later found, aurochs had been the first creatures of the wild to be tamed. The generations thereafter had been selected to be placid as well as productive. What he was seeing, looking down from the loft, was a sample of that transition nudged along for millennia by drovers.

If the young student stared at any of the domestic animals long enough, they began to look strange. He later saw that as a two-way course, like a trick yin-yang image - familiar to strange, strange back to familiar. Burrowing and scurrying insects also flipped, none stranger than the millipedes. Even in the family collie, Cap, the gentlest of dogs and his personal friend, he remembered seeing hints of wolf. Cap hadn't forgotten how to show his teeth and snarl, going in an instant from the familiar pet to a creature of the wild and then settling right back into Cap.

Biking on a county road, the 12-year-old McLaurin was chased and bitten by another wolf pack descendant, tamed not quite enough. To the earlier student, Cap had been just Cap, and even to the 12-year-old the bike chaser was only a momentary threat, not a place holder in the extended canine story.

Watching his father loop the tail of a cow into a twist to usher it up a ramp and into a trailer taught him a lesson in human mastery. Many of the things his father did, from tillage to transport, looked masterful to his son. Grownups disturbed plants, animals, and children, clipped the wool of sheep, milked cows, cut thistles, painted houses and barns, and downed trees. They pushed, pulled, and chopped nature into conformity. That makeover had been typical of many centuries of the continent's transformation, and of a span several times longer in Africa and Eurasia. For the most part, the unruly side of the acreage lay outside the fences in the wooded hills and sprawled along a turbulent river that marked the southern boundary. The fence around a lower field separated the open, groomed area from an assortment of trees, weeds, wild grass, and brush - prescribed on one side, unmastered on the other.

What a story that fence told him later! In human terms, a couple hundred thousand years of foraging on the wild side, a few millennia of cultivation on the groomed side. A lane along the fence marking the division provided the path he sometimes took on his wanderings. Some of his excursions carried him into the bordering wild - not very far, but enough. With the exception of gophers underground, birds above, and insects within, the fields were exclusively for crops drilled in rows. A little of the wild crossed through and dug under the fence; rodents and other pests, for instance, tunneled under the garden, scurried about the barn, and worked their way into the walls of the house where they rustled at night. Outside the cultivated areas, nothing much lined up except the succession of days and seasons. Creatures of all sizes up to the rare puma were on the prowl.

Bears weren't far off in the coastal mountains, but they never came in sight within the fields or bordering woods, as the coyotes, deer, and curious herds of coastal elk did.

In chorusing coyotes on moonlit rambles through Cap's scent range, McLaurin heard the tame and the wild making edgy contact. The prowlers remained on the path nature set for them. The grumbling Cap returned to his slumber with only a hint of the wild twitching in his sleep.

Other quarrels between tame and wild were less noisy – cats stalking mice, farmers attacking brush and thistles with sharp edges, gophers poking heads up to pull spinach underground, black ants fighting red ones and herding aphids in tiny versions of dairy farms. They came in squirming, stalking, flying, swimming, crawling, gnawing numbers and stayed mostly hidden, unless they were airborne. Midges, fleas, and mosquitos had people on their lists of prey. The wild things were what the cultivators called pests, weeds, and brush. These, the hills and mountains, the sun, and stars beckoned into the distance without his realizing it. He didn't go where they directed for many more years, but they, meanwhile, were filling his inner encyclopedia with the unruly and the ruled.

Despite being short on ideas at that age, he did have questions. Take a typical spring day. Everyone except him was busy six or seven days a week. To kill time while his sister was at school, he walked the lane past the lower fields to the river. Some of the pasture was a transition area, where knee-high grass was of a kind improved for grazing, sowed broadcast rather than by plowing, discing, and drilling. Closer to the river, the twangy song of red-wing blackbirds found a place among the raucous calls of crows. He hadn't yet thought of medleys, but where sights, sounds, and smells mixed, that's what they became - not *networks,* more like *miscellanies.* What remained in confusion was also a *hodgepodge* or *mishmash*, words he liked the moment he heard them. *Melange* was another he came upon later. He had watched his mother at work at a pharmacy using a balancing scale, a small scoop, and bottles of potions and powders taken from a shelf. What she put together were *formulas* or *prescriptions*. Gauged ingredients and portions in the kitchen were *recipes*. The more exacting the measurements, the more scientific the brains, hands, and tools had to be. Measurements. Scales. They were milling about looking for lodging in the young brain.

Measuring and bottling were mastery at work on a scale of grams and ounces. In contrast to forking hay for cows and scooping oats for the horses, that was *meticulous* (a word his mother used now and then, a polite form of *picky*). Being picky meant keeping things apart and in their places, not heaped up. He himself was being taught to be more meticulous, less a two-legged animal. That struck him as unnecessary, but he loved the words that came with it, a sort of miscellany of its own until someone lined them up and marched them as instructions. It wasn't until high school Latin that he would come across the marvelous *struere*, instruct, construct, structure, destruction, putting something together or tearing it apart, from the bones that held him together to the complete thoughts of sentences. A scurrying or flying insect? It had structure despite its frailty. On a spinning planet orbiting a star the busy insects hatched, grew, died, and disintegrated.

Developed and then lost their structure. Everything did, it seemed.

Down by the river a light breeze sent rustling noises moving through the alders and ash trees.

Close up, the rock-stirred river overpowered other sounds - except a distant calf bawling an off- stage solo. The calf's plight was all that mattered to it, and nothing that mattered to the blackbirds and crows. At the time he'd had no words to describe such a patchwork, only a few popup questions. The rocky river course would later return to him as if out of deep time. Intersecting courses were like it, or like interfering ripples on a pond. How was it that light traveled through sounds without disturbing them? It came from far off yet was always there, seemingly without moving. It had no weight, yet had a presence. Some things fit snugly together like spears of grass pulled from their sheaths. Mostly, each thing went its own way until it encountered something else. Toward the northeas,t as a showery mist drifted in from the southwest, a rainbow formed, the colors like voices in harmony. It cast an eerie glow over the field. Parts that belonged together like that formed a *spectrum*, a word he had recently snagged from his sister on its way by. It too, he eventually found, had taken an interesting winding path down through the ages in the company of spectral, specter, spectacle, speculation, and eventually spectrology, all on the lookout, ghostly, watchful, suspicious.

He retraced his path at a drifting pace, like thistledown in a faint stir of air. His sister wouldn't be home from school for several hours, and his father was busy in one of the barns with the hired man, Ted, docking lamb tails, nudging timid, half-wild lambs a step closer to the market. He came to a pen of ewes and lambs milling about and sat on the top board of a wooden fence to see if any of those he knew were there. They looked too much alike to tell. Inside the barn, Ted caught a lamb by a hind leg and carried it to a corner where Dad slipped a stout rubber band over its tail with tongs. The band would cut off circulation and, in a few days, the excess wild tail would drop off and leave a domesticated stub. Without knowing what was happening to them, the lambs were getting neater, more meticulous. Wanting to be of help, McLaurin climbed down among the sheep, and, after a grab that came up empty caught an un-banded lamb. When he lifted it in his arms, its kinky wool felt stiff. Its legs dangled. Its eyes bulged. A touch of the wild made it anxious. Each lamb took only a few seconds to pass from its mother's exclusive care partly to the farm's. Knowing the one he held would be sold when it reached about 80 pounds didn't change the pleasure he felt in handling it. He held it longer than he needed to, and it became more contented in his arms.

Finding himself an intruder, he left the barn and went up on the hill in the back of the house to one of his favorite haunts, a spring nestled among oaks and maples. The water was captured in a wooden cistern and piped to the buildings and garden below. The light green oak leaves would soon grow darker. The dappled light and the scattered objects were another kind of medley, gently wild like the spring captured in the cistern. In among the mixes, he had no difficulty understanding parasites like mistletoe. All around the farm one thing feasted on another, fleas on pigs, turkey buzzards on carcasses, mistletoe on oaks. He had been stung by mosquitos and bees and knew that kind of intrusion. He also knew something about mold, boring insects, and worms in ears of corn. Adults took such things in stride and scarcely took notice of them. A large mushroom-like growth on damp wood demanded poking and prodding to see how it responded. Spongy things weren't quite all there, it seemed to him. They were unfeeling and had no sap, blood, or bone. The oaks hosted galls, insects, and mushrooms as well as moss. He didn't yet know of tapeworms and bacteria or much about predator insects except for the mosquitoes and warring ants. An ant carrying a nibbled off leaf was a many legged harvester and a thief.

When he lifted the cover of the cistern and looked inside, the tank was brimming with water and had striders – pond skaters – floating on the surface. He waved his hand above them and made them scatter. Moving sideways, they soon drifted back into place. Each time he waved, they cleared a space and returned. That was a little like taming them, but he was only toying with them. When his dad and Ted trained something like a colt, they had a purpose. The colt would become a horse no matter what they did, but making it a trained, obedient horse took time and patience. Lifting the lid had revealed a water world not much like the river's. The trees mirrored in the tank looked strange there. He knew that was a trick of light, but that didn't change his sense that still water was another world. The surface reflections had a way of mingling - not with the water, not even really on it, but hovering within it somehow without touching it. The water striders looked at home there.

The farm had many small pockets of mixed populations like that, and as he knew from car rides, the valleys and hills did as well. It occurred to him that every place he had gone this morning he had changed. Birds fell silent or fluttered away when he approached. The cistern reflected his image, and the Jesus bugs avoided him. The things that responded to him had none of the mystery of the mirroring cistern. They were just what they were. A musty smell rose from moldy leaves and added to what was in the air. Whatever smells were, they came with moisture. He imagined a bird making its way through them and through sounds without disturbing them.

Invisible things must be letting foreign bodies pass through and resuming what they'd been doing, the way dust motes in a beam of light swirled around a waved hand and closed behind it. He was a small part of the stir.

The April drizzle had stopped, and the sun appeared between billowing white clouds. He perked up. Though he had no notion what sun rays were, something inside him responded to them. Under the trees around the cistern the colors brightened and dimmed and brightened again. When he left the spring and the woods, the shadow that went with him jerked across the rough ground like a second self, a virtual self he would decide later about such things. In the back of his mind, he had been working toward an abandoned rock quarry that because of fossil imprints in the rock was sometimes his destination. His arrival sent gray ground squirrels scampering. He held it against marauders–squirrels, gophers, deer, rabbits, chomping insects–that they ate crops and gardens. Let them stick to whatever it was they ate before the planters arrived. He lobbed rocks at the holes into which the squirrels had dived.

He poked around the inner wall at the deepest point and, with a sharp-edged rock, scraped a spot already scored by something. He could make out leaf and clamshell imprints. How could sea creatures have been this far inland? Trips to the beach took over an hour through the hills in the family car. A colony of clams must have lived here once, but how could these hills have been under water? When he reached out and touched the rough surface of the rock, feeling the ribbed shells only made them stranger. It would be many years before he visited the eroding cliffs of the John Day fossil beds. When he did, they would bring back the memory of the imprints. In that remembrance the Cenozoic became personal. It was *his* time but spectral among his other memories. The past receded in stages, without a decisive division between things truly there and memory's dim version of them. The images not quite touching the cistern's water, yet somehow within it were similar hovering things. *Residue* was another word he had yet to learn, but relics and traces left behind were part of what something long past had left scattered about. In meandering he merely encountered such things in the company of one another. History lined them up and made a sequence of them.

He turned toward the house, noting the flight of a hawk into the woods where its nest must be. When his mother was at work in town, Mrs. Hamacher came from somewhere up the valley to fix lunch. While he ate, she added to the miscellany by telling him a story in a German accent.

That was one of the several mixes in the valley: Japanese labor and owners in the onion flats near town; English and Irish descendants of those who had come west, and the offspring of those locally called Oakies and Arkies; a few Hispanics; Indians on the reservations and about town – people, too, were a hodgepodge. After lunch, he tagged around behind the Swede, Ted, for a while. The span between lunch and dinner seemed endless, but eventually the school bus came, and his sister climbed down from it. Between its arrival and dinner, they had over two hours for chores and play, but something more interesting came their way. Ted waved and yelled to them from the barnyard *round up time!*

To bring the dairy herd up from the riverside pasture, Ted sometimes galloped them three on a horse. Nothing seemed wilder than Patchy at full gallop, her tail flying out behind. They considered Ted a daredevil, half wild himself, at that job. He was undaunted by butchering a hog, dehorning a cow, making a jumpy horse behave, or felling a tree at exactly the right angle to avoid a fence. He wasn't cautious now either. With the two of them hanging on behind, he took off over the rough ground at a gallop. Crossing the line between the domestic and the wild, the dappled mare took command despite the bit in her mouth and load on her back. In herding the cows toward the barn at a walk, she relaxed. They all did. The cows were obedient. Patchy was herself again. Ted returned to the amiable Swede with a grin and a joke. They were an ensemble, people, horse, cows. *Ensemble* learned later: another bunch made of assembled parts. A flexible structure. A *composite*.

After McLaurin's mother had brought groceries home to go with canned goods from the pantry and set about preparing dinner, the family gathered. The sun was setting and the medleys of the day were giving way to those of the evening. They included crickets and the bleating of lambs that hadn't yet relocated their mothers. The kitchen was the center of the family's indoor world, and an electric range the heart of the kitchen. Mixtures there were like the chemicals his mother put together at the drugstore. A red-hot burner, like a tamed sun, was the life of the range. The refrigerator, with its well stocked shelves, was their second - an improved replacement. He remembered Ted and Dad moving it up the porch steps, plugging it in, and listening to the purring sound it made. They had looked at one another with a satisfaction similar to his dad's when the new tractor arrived, or not quite new in that case but better than the lug wheeled one rusting in a corner of the barnyard. Inside the refrigerator, he knew, was another mix, an edible miscellany in containers. Like it, the pantry was set aside for organized storage, in that case for canned fruit on one shelf, canned vegetables on another, and canned pork sharing a shelf with empty pints and quarts.

The books that McLaurin's sister brought home from school were different from anything he came across on his rambles or knew in the kitchen and pantry. He couldn't read very well, only what she had taught him, but the pictures helped. What appeared in books was meant to be there, no more a casual mix than what went into a prescription or a saucepan. Farms had boundaries. Kitchens had methods and a place for everything. A book was both itself and more. Print *sun* in one and the word raised an image of the object. Unlike reflections in a cistern, that was a strictly mental image. *A long way* meant over a mile. *High* meant out of reach. *Hard* was the feel of a rock, *cold* the feel of the river. Dust motes came *small* and *scattered*, lamb's wool *kinky* and *stiff*. Such things fell into place better on the page than they did about the acreage. Chapters were verbal cupboards. Things outside books were real. Inside they became images and ideas. In the reading, each word waited its turn before taking its place. Words in combinations had structure. Books cultivated and weeded them.

"Let's go saddle Taffy," his sister proposed after dinner. With the hard-to-pull calf nominated, Cap wouldn't be made into a small horse pulling the little red wagon. A *virtual* horse. One thing playing at being another. Shapes etched in rock and things named in books or reflected in cisterns weren't quite real. When relativity was introduced in his advanced curriculum, he had no trouble multiplying its conditions. Everything came in relative sizes, velocities, and in presences from phantom to full. Light came in relative brightness. States of mind came more or less certain. Combinations collided or cooperated in a ceaseless stir. His own moods played through them like mottled light on the cistern. Words in books referred to things not there as if they, too, could be relatively real, not quite there but more so than fantasies. The books were methodical, made of orderly lines sort of like pantries for idea storage. Words arranged thoughts like labels on a container, *peaches* on a canned jar of them. Tagging, labeling, assigning: that's what words did, the way a name for a species pinned it like a leaf or a butterfly to a display board.

With each year those wandering days came and went in fainter recollection until they, too. thinned out and seemed less than real. An old chaos of the sun, Wallace Stevens said of the mingled things the passing days brought. The older McLaurin didn't want to make too much of the roaming and milling about, but the miscellany had been noisy in the rock-troubled river and in the quarreling dog and coyotes. That had been a fine April day with a rainbow. In the dead of winter, Stevens (in "No Possum, No Sop, No Taters") thought of a frozen field where "The leaves are dry./ Bad is final in this light" (247). In these parts, the dead of winter came with bare tree limbs and cold damp winds, occasionally with snow, each season on a carousel with the others making what was past, or yet to come, less forceful than what was present now. Sequences, melanges, mishmashes, hodgepodges in motion were roughhewn, robust in behavior, constantly astir, yet on clocks and calendars fixed and predictable. Left to unrecorded impressions they became phantoms. Everything not close by announced itself first from a distance and, after passing, weakened and vanished like memory's own Doppler effect.

It wasn't only the wild animals and the miscellany but the people McLaurin came to see he hadn't understood, not really. He could barely remember those who weren't family - except for Ted, with the challenging look in his eyes, fiery one minute, relaxed and jovial the next. In memory the racing mare, neck outstretched, tail flying, accompanied him as if in a living panorama with a background of cows in the pasture and a mingling of crows in the air. Patchy galloping. Patchy walking behind the cows. Ted twisted around and grinning at his two relieved passengers. (You weren't scared now, were you?) On the ground, insects scurried over broken ground, mountainous to them. In the sky, birds were jostled by the wind. In the shadows of the riverside woods, ancient eyes kept watch. Kind of spooky, he decided later, these hints of more than he could see or name. At night, off among the moon and stars; in the day, close by in air moving inland from the ocean, or coming off the deserts and down through the mountains.

Coming laden with hints of things he couldn't fathom.

CHAPTER FIVE

ENLIGHTENMENT COMES SOLAR POWERED

The Argument

Going only where natural history permits frees naturalism from ritual and magic as well as from dubious beliefs. To some that has seemed a loss, but what it observes is demonstrably real and doesn't generate the collective fervor of some groupthink. Natural history is more extensive and varied than what is usually substituted for it. The sun of McLaurin's youth wasn't the lead-in to deep history it would eventually become, but he was unknowingly collecting what it would include, along with some of its literary, artistic, scientific, and philosophic contexts.

< >

There is a master of mud. The shaft of light
Falling, far off, from sky to land, That is he–

The peach-bud maker, The mud master,
The master of the mind.
(Wallace Stevens, "Mud Master")

The Dawning

Sun lore historically came with anthropomorphic inventions that differed from culture to culture. It is still nowhere near complete, but the launching of the Solar Dynamics Observatory (SDO) on February 11th, 2010 accelerated progress, and what it records is selectively available on its website (sdo.gsfc.nasa.gov). Even closeup observation won't eliminate popular level sun lore, of course, but for the most part the sun gods are gone. One major error typical of placing humans at the center of the universe was corrected four centuries ago by a handful of astronomers who had the planet orbit the sun rather than the sun the planet. Nicolas Copernicus (1473-1543) wasn't the first to argue for that revolution, but, reinforced by subsequent astronomy, his contribution was the most persuasive up to that point in getting the sun, earth, moon, and the other planets placed in concept and representation as they are in space. The mixed promise and problems of that led Galileo (1564-1642) to endorse the heliocentric revolution, set it aside, and endorse it again.

Aided by a magnifying lens turned upward, his contribution to the altered model was predictably unacceptable to those who maintained that a stationary, always-centered sun couldn't have been commanded to stand still. Given the assumption that what scripture said had been transmitted directly from the Holy Spirit, and that the science wasn't decisive, they had a point. Planet and moon movements worked out about the same either way. It was only bit by bit that Aristotle, Ptolemy, and doctrine based on scripture were corrected.

The former way of getting from mankind out to the distant heavens was simply to place the planet at the center of everything and have heavenly bodies whirl around it. That was how the biblical creation had it, and that's what it looked like every night and day. Galileo's way of making the case for a heliocentric system without sacrificing either the privileged human position or the key biblical text in question (*Joshua* 10:12-14) was to insist on a principle no one in his day, or very few, questioned: "The holy Scripture and nature both equally derive from the divine Word, the former as the dictation of the Holy Spirit, the latter as the most obedient executrix of God's commands" (Letter to Castelli, 1613). In a letter to Grand Duchess Christine (1615), he put that in the more elaborate manner of the times. No one seemed inclined to go into the implications of widespread suffering and mortality if nature was a most obedient executrix. What Eve ate one day explained everything, minus a few celestial quirks. No one yet knew of course of the debris and loose materials throughout the cosmos. Ultimately, he couldn't avoid the contradiction his critics insisted on between a sun-centered system and a text that had the sun stop in place and stand still at the request of the chosen people. Hence, for most, the sun and other stars continued to orbit the planet for a while longer. No one was in a position to calculate the immense distance that points of light orbiting the planet would have to travel - trillions of miles per hour out on the circumference. Lost in the debate was the reason the sun had been commanded to stand still to begin with, namely to finish the slaughter of every man, woman, and child in still another city under siege. Ordering the extermination of non-cult abominations was the primary function of most appearances of Yahweh claimed by the Hebrew prophets and priests of several eras.

The debate over scripture and its relation to nature broadened in the next three centuries to include other ways in which phenomena ran counter to revered texts. Empiricist-minded philosophers well into the 19th century were still trying to credit both the facts and the texts, despite the evidence of a vastly extended earth history and the number of species. Either the latter had to be few enough to load onto a homemade boat, or they couldn't be reconciled with the story of the flood which, in that case, portrayed the supreme power of the universe killing all land creatures except the one boatload. That was an especially gruesome act of wrath against "all flesh, wherein is the breath of life" (*Genesis* 6.17).

Although not many put the shift to observation and theory nostalgically any longer, Franceso Panese offers a trailing lament in "The Accursed Part of Scientific Iconography," a well illustrated essay in a collection of traditional bestiary imagery, depicting human and other animal shapes sporting leafage, fish tails, horns, and assorted lumps (*Visual Cultures of Science*, Luc Pauwels, ed., 2006). One point such a collection makes, without having to state it, is that age-old fantasies are still available even though they aren't any longer taken literally. As Daston illustrates in *Wonders and the Order of Nature, 1150-1750* (1998), the old icons and the stories that came with them have actually become more available in publications. Add to those numerous internet sources.

Thanks to the accompanying scholarship and commentary, they have also become more understandable in psychological, political, and cultural terms. What Marco Polo brought back from China and India wasn't disseminated by a printing press or contextualized by scholars as the publications became over time. Thanks to archaeological excavations, the record of cultural artifacts overall – not merely those of western traditions – has also grown more ample. Dynasties in the East paralleled those in the Middle East and West in the invention of events of wonder, and the use of astrology to connect distant stars to people. In plain sight and close to home, the forests and deep seas contain both a variety and a strangeness exceeding that of invented creatures, even where that inventory includes all cultures and their myths. Certainly, the mess of facts after the 18th century encyclopedias was no longer the paltry meal of pottage Esau consumed in one well-known lament over a lost birthright. Naturalists catalogued more of everything, grander and more terrible than anyone had dreamed existed, in a few centuries. Baal and Thor weren't needed in person to contribute drama to storms that had all the thunder and lightning anyone needed on their own.

The names of the days of the week didn't need an actual Thunor or phantom figures behind Sunday, the moon's day or Saturn's Saturday. Tiu, Woden, and Freya had always kept out of sight while the workweek proceeded, much as the Holy Spirit kept aloof from storms raging and wolf packs prowling.

The case for myths and miracles isn't helped by making measurements mainly how empiricism conceives of things, as Panese seems to do in commenting on the "homogeneity of measurement" and concluding that "our own contemporary perspective, our way of looking at the world, is still so marked by this [homogeneous measurement] . . . that we often need to make a conscious effort to efface this perspective in order to consider it as just one particular way of rendering credible the visible" (*The Rhetorical Turn,* Simons, ed. 83). It's an apt reminder, but would present a problem only if measuring was all the naturalist and empiricist method did, and it hadn't proceeded into the marvels of the micro and macro extremes. In the context of the scalar spectrum, what the measurements mainly do is allow the relative placement of one thing in the company of selected other things and of the total. Measurements are relational and thus, along with velocity and vector, instrumental to placement. We can't get things sized, timed, and placed properly without knowing where they fit and how they are related. Meanwhile, the measurements don't remove the pouch of the kangaroo, dull the conscious stare of the octopus, or remove the smell of the polecat.

Questioning "whether the ideals of order and of measurement proper to the scientific approach" were rooted wholly "in the puritan ethics of the gentle-men-philosophers who first posited them," associated with "English Protestantism" (*Visual Cultures*, 85) leads to no conclusions one way or another. The fields of interest in an era are certainly related, but not necessarily in a straightforward way. Empiricism's adherence to observation has nothing in common with Protestantism specifically, except in some cases stripping away excrescence and colorful ritual. In England, the Puritans weren't among the leading scientists and philosophers - although some Protestants were. If any puritanical spirit cropped up in the controversy over the heliocentric model, it wasn't in what Galileo or Kepler reported about what they were seeing, but in those committed to keeping mankind the focal point of the universe. Protestants weren't more likely to follow the sky trackers and Newton than Catholics were, and those weren't the only two alternatives. New wonders were appearing in world travel and in what empiricism set out for public view in sketches and publications. Wonders around the globe were being mapped and placed. Being too curious about nature was frowned upon by several branches of science denial including, as we've seen, the flagrantly brilliant Swift and Pope in the next century - but that didn't prevent the compiling of field guides, maps, and encyclopedias.

Nor does a culture that encourages inquiry necessarily produce results that reflect whatever set of ideas and conventions is dominant at the time. Just the opposite is the case where discouraging inquiry favors a ruling hierarchy and curiosity turns up observations unfavorable to those in power

- the ones who put Galileo under house arrest, for instance. Where the pursuit of learning doesn't require social stature, it becomes too democratic for those who prefer power to knowledge. Calling accuracy just one way of handling what is visible is fine, it *is* just one way, but equating mermaids with dolphins isn't a comparable way nor is humanizing natural phenomena. No one needs a sun god to continue harvesting discarded solar byproducts (light and heat) nor a Noah story to moralize a heavy downpour. More convincing is John Lyne's conclusion that "externalist critiques of science which would, in effect, perform a reduction of science to ideology, simply boil away what is distinctive and powerful about science" (*Rhetoric and Community,* ed J. Michael Hogan, 271). Conversely, as those in literature know, Spenser's *Faerie Queene* is no less revealing for being fanciful, nor do wonders lose their fascination in being classified as fiction rather than history. Modern readers don't fall in line with everything in Spenser, Sidney, Shakespeare, Jonson, the metaphysicals, and Milton, but that doesn't remove those writers from anthologies or diminish the profit and pleasure we get from reading them.

In any case, nature, subject to scrutiny with or without accompanying marvels, is remarkable in its own right, and incredible in strangeness as well as in extent. The revealing question about the myths that persist is 'if this gets reified, who benefits?' *Reified* is multiplied exponentially in *deified* under the anthropomorphic principle of making humans into gods. By that simple expedient, mankind becomes connected to everything that exists. A national or imperialist power isn't the only answer to who benefits from the increased power and authority, but it had already become a common one when texts were first inscribed in stone. Social classes and venerable institutions become other answers in due course. Ancient Israel was only one of many nations whose heritage was mostly legend and not nearly as important to the universe as the local civil and religious authorities maintained. When all other forms of anthropomorphism are gone, claiming a powerful invisible connection to the cosmos for a nation may remain precisely because it strengthens social bonds and governance. Convocations and other forms of grouping use choral recitations to reinforce the public function of that fiction. Such recitations are more likely to be addressed to a sun god than to a heat and light emitting star, one of gazillions of nuclear fusion power plants. The unavoidable fact remains that nuclear fusion is real and sun gods are not.

Stevens captures that in "Mud Master" in having a shaft of light, rather than a divine revelation, fall to land from the sky and having it master mud, not a chosen people. That it also masters the mind means any mind focused on mud and peaches as well as on what actually sent the heat and light, unknown to Stevens, discards from a thermonuclear process that each stage of higher element generation throws off. In generating green growth and crops, the sun perforce attracts the attention of biological creatures. Sun plus mud plus biology generate the harvests to which creatures apply themselves. The fruits of solar output are real, not metaphoric - except in literature.

Works and Days in Retrospect

I want to pause with that reminder of the nearest stellar power plant and pick up the test case again as a model for taking in the scalar spectrum bit by bit. Stellar advances in the offing require a major step for the maturing McLaurin, as in most possessed initially only of a personal sun beating down on him. It is a long way from that to understanding the atomic matter invariables, especially the realization that nothing lies outside the atomic matter continuum. I won't have him take detours through any of the popular myths, since very few any longer put Woden or Freya in charge of selected days of the week or imagine a supernatural intelligence managing the sun. That a good many still personify some fraction of nature, however, testifies to the durability of regional beliefs, both the pressures of groupthink and the anthropic temptation to moralize the universe evident in them. We'll resume with McLaurin as the template example at a stage at which he enters the work force and begins applying concepts acquired in secondary and university levels of education to what he sees around him. Connecting this and that in one's experience to the continuum is a many-sided affair.

To nearly everyone, the sun is intimate and plain as day, and yet also an atomic phenomenon, somewhat bewildering. McLaurin cursed and blessed it in turn depending on how it came to him, too hot on an August day, too weak in a frigid January. That was personal and unavoidable for anyone outdoors a good deal, employed as McLaurin was as a harvester earning school funds. On summer workdays the sun took the damp off mornings and let the harvest of berries, beans, barley, oats, wheat, apples, plums, and peaches proceed. The crops had plant histories as well as organic responses to light and heat, but McLaurin knew little about those at the time and had only the vaguest notion of how photosynthesis works.

Heeding a new day's call in a timely manner is more important in rural settings than in urban ones. The latter are keyed to the rising sun differently than where a lingering morning damp or drizzle reschedules a day's work. Urban time is as much clock time as sun time - whether or not the sun is out and shining, papers get shuffled and rivets riveted. For everyone, urban and rural alike, however, the nearest star is a task master and regulator. What foragers had sought was at hand, and they gathered without planting.

Urbanites can also, hunting up supermarkets and gathering cartloads of produce from shelves and freezer bins. In a food producing setting, however, atmospheric conditions, especially temperature and humidity, set the work schedule.

The Pacific Northwest, McLaurin's particular setting, was green much of the year from spring and early summer showers, but turned golden in the fields and pastures at harvest time and gray in open areas in late fall and winter. For everyone, that is the most obvious manufactory of solar power, its mechanics dictating the buildup/breakdown phases of the vegetable kingdom and thus the feeding and shelter-seeking habits of the animal kingdom.

Much else is solar produced as well, but not so brushed into surface tints. The planet and sun together make a dynamic pair not beholden to other stars and satellites. The working mechanics of the celestial bodies aren't anything a harvester needs to know in leaving the house and checking the sky. Not just McLaurin but nearly everyone goes about a day's work without much thought of these mechanics and atmospheric chemistry. The seasonal patterns are obvious, but to find what turns the mechanism requires more instruction than most students ever encounter. What matters for a harvester is the sun drying out the morning damp that, even in July and August, can thicken into fog, and hang droplets on stalks and leaves.

As to the tools and equipment the harvesters use, following formats of their own are machinery invention, manufacture, and marketing, and, on the consumer side, repair and replacement. The course of the devices they use from design to rust turn many times slower than the seasons, but faster than an average human life cycle. As an economic mode of production, supply and demand set the amounts. Though they are economy and technology driven, it is the sun that powers them. The machines are designed to handle what the sun, soil, and moisture produce. Timing in and around internal combustion engines includes the extracting of oil-based energy, derived from the sun millions of years ago and stored underground, refined and exploded in exact amounts at millisecond intervals. In buying tractors, trucks, and combines, planters and harvesters take delivery of more than the machines. They buy into a way of life linked to blueprint invention, oil fields, coal mines, steel mills, and banking. They join an engineering and financial network that includes labor unions and training for skilled jobs, dozens of different ones. In getting synchronized, these draw on the planet's orbiting of the sun, its tilt, and daily rotation. Money goes on a circuit of its own - with corporations dropping off percentages to executives and workers on a yearly or hourly wage, to stockholders, and to iron workers and other metal and plastic industry technicians. That flow of funds imitates the flow of energy and, at several removes, is likewise sun-governed. The stock market generates weblike connections and concentrates into nodes of its own. A transferred dollar is like a watt of electricity or a calorie, a small unit hooked up with countless others. Currency is information in emblematic form, as symbolic of capital transactions as crowns are of monarchies. All these, too, are at several removes from solar power but indirectly moved and timed by the sun. Heat stirs everything that moves, and what little of it isn't solar in origin in planting and harvesting comes from inner planetary storage of a former star's disintegration. Directly or indirectly, photosynthesis in the barley becoming beer means earnings for breweries and workers. Long-stored fossil fuel consumed in planting and harvesting goes into current profit and loss calculations and the purchase of loaves and six packs.

That is, broadly speaking, the multifaceted process that assigned McLaurin his primary function in the system. The willingness of earlier land owners to enlarge acreage and join the mechanized revolution elevated not only their own claims to middle class standing but those of their suppliers and technicians. Staffs in sales and repairs found a place in turn. The specialization that supplied the skilled labor grew uncertain with new inventions, but knowing very little about that didn't keep field hands from being enablers and technicians at the levers. Their contributions were synchronized with thousands of others. From rubber tires to aluminum outer shells, every labor-assist machine the growers used had passed along an assembly line, with production speed keyed to hand speed and wages set to clock time. That, too, of course, was keyed to the sun and matched to what it grew.

The assembly line McLaurin joined as a harvester was staffed by technicians, their movements adjusted to the metals, wood, plastics, and fibers. Whatever the value of the raw materials on which they worked, – and it wasn't the zero of Marxist theory, explained in one of his classes – it was increased by what they put into it. The most rarified stations in the assembly line were staffed by chemists and mathematicians, whose manipulation of wood, metals, and plastics was getting along well on space age materials. The most tedious and nerve-wracking parts were the conveyor belt tasks depicted in Charlie Chaplin's *Modern Times.* That early toy film with the serious message was one of McLaurin's favorites. In it, in animated cartoon fashion, a limp human body gets wrapped around gears like a chain in cogs. Those who drive harvest machinery understand that, and liken it to their own personal hours at the wheels and levers.

Tanking stored solar energy and exploding it in cylinders is more efficient than stoking horses with hay and grain, if more costly when design, production, fueling, recycling, and pollution are factored in. McLaurin, little by little, becomes aware of what being part of that synchronized network means, all parts of it solar powered.

Household furnishings and appliances, too, follow irregular rhythms of innovation, manufacturing, use, and discard. Manmade things, from engines to Rube Goldberg contraptions, are keyed to RPMs, some of them in the thousands. After a time spent in limbo, discarded materials get recycled. The junking of one implement and acquisition of another has as much to do with modernization as with wear and tear. Engineers and consumers speed up the buildup/breakdown cycles that the materials would follow on their own eventually. Indoor equipment, too, gives out or gives way to better models in a few years - just as the wood stoves and hand-turned eggbeaters of previous generations yielded to electric ranges and whirling electric mixers. Some time ago, wood-burning heaters in the living room gave way to the diesel furnaces of central heating. Over a few decades, the use of sun-produced energy has expanded and been revolutionized with probably not one in a thousand placing it among similar cosmic mechanisms. Chaplin on the assembly line and harvesters at steering wheels and levers are half robotic - that much any machine operator can see without instruction.

McLaurin could easily understand the sun as a taskmaster, ruling over those summoned at daybreak and dismissed at the dimming of the day. Losing sunlight is ambivalent to creatures of warmth and light like himself. That is as much subjective as objective, as it was long before the first planting and harvesting. Sunset and evening bring the harvester rest and a sense of emptiness. "I need you at the dimming of the day" a Richard Thompson song confesses to a hoped-for confidant. All creatures of the sun, not just people, deep down feel the disappearance of light. They need it as much as the green growth does. No more than botanical mechanisms can biological ones adjust to the dark as easily as they do to the light. One or the other dominates, and so round and round it goes, gears within gears, cycles within cycles. Together, the farming, industrial, and social mechanisms have as many moving parts as a Swiss clock shop.

The Slant of Light on a Harvester's Workday

When the sun was in sight, the tallies the work force read from its progress were like splits to a runner. Streaks of summer light showed in the east by 5:00 AM and lasted in the west until late evening. While the mud of winter and early spring hung on, the machines stayed out of the fields, but when the mud master came, with strength they went forth again. In the harvest operations of the northern hemisphere, dust and pollen flew until, for McLaurin, the long daylight hours of July and August seemed too much of a good thing. As the machines rattled and groaned, the bodies and brains in charge of them grew tired and rebelled. In early, July the mowing of crimson clover, one of his assigned jobs, had to get underway before daybreak. Otherwise, once the sun had removed the dew, a pounding sickle would shatter the seed heads and spill their contents on the ground. Sun plus sickle oscillation plus crop dictated when his alarm clock sounded. A harvester of old might have waited to hear a rooster crow and worked to the rhythm of a swinging scythe, but that was no longer the case.

The combine in use on crops across the grain belt was so named because it combined cutting, thrashing, separating, and sacking or grain-bin accumulation, synchronizing what had formerly been done separately. The sounds and shaking of the pulverizing machine McLaurin later saw as symptomatic. The entire operation was solar governed. In due time nature knocked plants apart on its own, but harvest machines speeded up the process, saving one part of what was separated for further processing and discarding the rest in, unknown to McLaurin, a process similar to the sun's harvesting of helium out of hydrogen. The combine inside had shaking straw racks, and an elevator that conveyed the seed in metal cups up and over a sprocket and spilled it down a chute into a burlap sack or a bin. That was the most exactly synchronized operation McLaurin knew personally, since he wasn't familiar with high production factories and assembly lines outside of movies and documentaries. A practiced operator who listened closely could hear distinct parts at work; the engine, pounding sickles, whirling cylinders, rattling chains, and elevator cups in their chutes. It took another level of detection entirely to fathom the thermonuclear processes of atomic matter, simultaneously building up new composites and breaking down their sources.

McLaurin pictured Chaplin caught up in the whirl and clatter, wound around the cylinder, shaken by the straw racks, and blown out onto the field where he lay battered and bewildered on the ground. In its devouring of what photosynthesis had produced – the green growth harvesting of solar discards – the operation did have a vaguely animal presence. The sickle sawing back and forth (teeth) severed the stalks. The paddle reel (the tongue) knocked them onto the conveyor belt (the throat). The machine spun, knocked, shook, and hammered its way around the field, its wheels leaving temporary imprints in the dry dirt. In separating what to sack and what to discard, the combine laid out rows of straw around fields or in strips across them that made a harvested field look geometric. The straw could be burned, but was usually plowed under or left to disintegrate on its own. Under the solar field commands dirt took in the fragments. In the muddy season it then dissolved them into its rich compound. That was the last stage of their breakdown in preparation for roots taking them in and feeding them into the green grown buildup of crops in the making.

Late afternoon sea breezes toughened the stalks as moisture won the next battle with the drying sun. In less completely mechanized eras, threshing crews with teams of horses had gathered shocks of grain as soon as the sun allowed. Crews on the ground were the cutters and conveyor belts of the operation. A family picture album recorded selected parts of that older method, and McLaurin saw other segments in period movies like those made of Thomas Hardy's novels. While stationary threshers remained in operation, a horse-drawn device equipped with wooden prongs out front, forerunner of the forklift, acted as a scoop, the job of the combine's header gathering its materials. When a load arrived at a stationary thresher, one crew member pitched it onto a platform and another into the maw. On a second platform, someone with manual dexterity was equipped with needle and twine and sewed the sacks. By McLaurin's day, the combine's grain tank and an unloading auger had taken over that job. The more direct and obvious products of the sun and human responses he understood even at the time. In a preceding generation's era, someone went out with a pitchfork to turn them damp side up before the sun had thoroughly dried the windrows. McLaurin's sampling of that job had come at about age nine or ten in a place where a stationary thresher, the only one left in the county and perhaps the state, had lingered past its time. From Dickens and others read to him in the evenings, he had an idea of what involuntary child labor was. His was entirely voluntary, albeit bribed by a small wage. As he soon discovered, taxing work went agonizingly slow. A fork inserted, pried up, and hoisted was as basic as a hand reaching out, grasping, and lifting. He eventually put that together with a foraging and harvesting legacy as ancient as hands became free for collecting and carrying in the first bipeds. The fork extended the arm and its tines sharpened the fingers. Some summer mornings were already uncomfortably warm by 8:00 AM, and the warmth made him lethargic.

By 9:00 AM he was listless. He had taken the measure of work in the sun, and it had taken his and found him wanting. Like green wheat, he too needed seasoning. His natural inclination was to sit on a hillside and watch the panorama play out, which was more or less what he did in collegiate and post collegiate days, putting more into understanding and less into doing. His harvest, by then, was often of data and ideas arranged in classified tiers under general topics.

Realizing that came several years later in library and class harvests. At the time he was mainly a participant. For several years, leased acreage in the coastal hills added to that of the home place and increased his workload.

The ride out and back along a rocky dirt road bracketed the workday. Through oak and fir boughs, it offered glimpses of hills and foraging deer, grouse, quail and, on occasion, a nocturnal predator: a coyote or an owl, running across or flying over the rutted road. These, too, were harvesters in a hunting and gathering phase. Some nocturnal animal hunts lasted into daylight and others began with it. From spiders to hawks everything alive was sun governed.

When a brief respite from work came at noon where lunch pails were in order, time skipped a beat. If the buzzing of sun-activated bees and wasps didn't put McLaurin to sleep, the crackle of bursting pods did. The mind sank in upon itself, and a half hour in the shade passed quickly. He later associated such interludes with things Mediterranean and Keats' Autumn Ode, but it was summer, not fall, and the 45th parallel, not Greece. He and his father spent those lunch breaks in a locust grove thinned to make room for a house, now gone except where the foundations were obscured by weeds. Time measured in decades was inscribed in tilted fence posts and a tumbledown barn with warped, sun-bleached boards. The decline of the homestead coincided roughly with a human life span, that too solar powered, though he couldn't have said how. At a time when the cost of machinery advised adding acreage, small remote places like that one had been abandoned as places to live. In a flaming hour, the ramshackle house had burned, but the barn that had kept it company was still there, leaning in a slow pantomime of decay. Almost a century, he calculated - that's how long it would take for the timbers to give up, bleached by the summer sun, soaked and softened in the months of dimmer light and dampness. Many generations of wood borers and nesting spiders would come and go, their cycles of growth and disintegration brief and their lives fragile. He later marveled at the synchronizing, beginning with the speed of light arriving, dropping down to the two miles an hour the combine traveled, the two or three century span of fir and oak, his own cycling through allotted days, everything turning until it didn't any longer. The units, wheels, gears, the plant and animal lives, turned partly under his management. What a harvester did was manipulated a phase of the growing, use, discard, and breakdown process.

While going round and round a field on a planet going around the sun, he sometimes passed the time by inventing riddles and taking imaginary forays. As the brain applied itself to the task at hand, the mind or some corner of it went off on its own. It turned over words to see what lay underneath and rummaged through sacks of ideas to see if any were worth keeping. Mind- bending exercises usually kicked in about the time adolescents acquired a liking for puns and teasing riddles. What preceded the beginning? Does the universe occupy an area? If so, what lies outside it? If space is nothing, how can it be said exist? If other universes exist, have they come one after another or all at once? Are there many of them, and, if so, are they at this moment spinning through predetermined cycles of accretion and disintegration? When one of them ends if it does, what happens to what it contains? Do any of the other stars serve as suns to forms of life? The brain developed and operated in the hum and buzz of life. When nothing was stirring, it virtually shut down like a windmill when the wind dies. But not the min; it claimed leisure as its own when reflex action gave way to reflection, using words to put some ideas together and critical reason to discard others.

He shared these meanderings with no one. In the fields, he was alone under the sun, aware of only a fraction of what was solar powered. It wasn't until years had gone more into understanding and less into doing that he found he'd been playing scales all along without knowing it.

Dust

One of the things McLaurin later reviewed would have seemed a most unlikely subject for analysis at the time: the disintegration phase of dust, dried by the sun and stirred aloft by solar powered winds. Thus, sent abroad, it and chaff landed gently and served their purpose seasonally. It later became a sample of celestial debris, gas, dust, and larger pieces broken loose. It was a very old notion and not exclusively biblical that dust was how people, too, started and ended. It wasn't as well recognized how far back and out into space that carried, with interstellar dust spread across a good deal of the cosmos. From the sun's station in the Milky Way, the line of vision in some directions was obscured by it. Is Earth dust a long way from cosmic dust? Yes, in amount and makeup, but similar as a stage of reduction. In variety, earth dust probably exceeded that of moons and other planets, since nowhere yet discovered put formerly living compounds into it. Settled dust buried what it covered under a dry layer that dampened at night, eventually getting rained or plowed into the soil. In the slanting rays of the setting sun, it became ghostly, like dusty celestial haloes. In the dark on the way toward subsidence and that dew-dampened state, it became information in the nose and throat. Its particles were immune to mutual attraction and repulsion, and therefore went where air currents directed. What all sizes of disintegration, large and small, shared was the disconnection of formerly combined pieces. In that respect, all discards are in limbo.

Harvesters and other work forces did their share of composite breaking. In rural settings around earth's crust, implements pulverized what needed to be broken down. Tractors, plows, and discs turned up insect lairs and worms, throwing them into turmoil. Sending noise ahead to announce a disturbance on its way, they assaulted snakes and flushed out birds. For McLaurin, in charge of a D4 Cat in late summer, nests of yellowjackets made discing an adventure.

Agricultural procedures disturbed a good many things in converting the wild into the domesticated and in breaking down compounds, turning over sod, pulverizing, sending particles abroad. For McLaurin, each trip around a particular hillside field ran a gauntlet of disturbed lifeforms. Storming angrily around the machine in defense of their territory, riled up yellowjackets were as vivid to him as impressions of childhood glory were to Wordsworth. As the clanking yellow machine passed, thinking it the source of their trouble, they swarmed out and wasted themselves stinging iron. Nothing in their evolution had prepared them for the iron age. Nothing similar had broken into their ancestral tunnels and turned private bowers inside out, to be cut up, harrowed into small bits, and some of it sent forth on the wind as dust.

In rural areas, September and October were devoted to that and to burning straw, leaving ash behind for soil enrichment. Some of the smoke and dust drifted off, but most of it settled nearby and, in the next cycle, turned into something green and growing. The nose and ears detected some of what the eyes missed; substances finer even than dust as floating, vaporous hints of smells. The smell of rotting apples clung close to the ground where bees crawled sluggishly. Fall and winter were the main seasons of vegetable decay and fermentation. When a late September day was neither summer nor fall, midday could scorch it back to August, thinning and scattering the remaining smells and raising twisted columns of dust in small cyclones. Nightfall came with a hint of winter. The seasons brought dissolution and renewal in many forms from dried grass to falling leaves. The main renewal phase was postponed until spring, after water, the great solvent, had blended elements into something worms could stir and roots absorb. Late August on the 45th parallel carried preludes of autumn.

By mid-October in the northern hemisphere, the sun that had overheated summer days became a blessing, and by the end of that month it appeared only faintly in the southern sky. After the plowing and harrowing came intermittent rains, some of them thunderous. The dust, by then, had become mud, and Argentina's mud was turning into dust. If plows and discs were to function a few weeks more before their winter hibernation, the sun would have to draw moisture from the ground - lest half a tractor disappear into the trench its churning wheels were digging.

Machinery designers put cleated tracks on selected models, but even with them by November outdoor jobs that needed fair weather had to be postponed. December and January portioned out clear skies stingily, and the weak sun came smothered in fog and mist. Outdoor dust was all but finished for the year. The sun left remnants of light in crooked lines beneath deciduous trees and in spots beneath evergreens. Even in its weakened state, it carried out its mapping and coordinating functions.

A month into the sun's return, February could herald spring, but the ground remained soggy and last year's grass lay downtrodden. Creatures who were as seasoned as McLaurin was becoming knew the promise would be broken. For those who liked melodrama, March and April mixed a stirring clash of elements with gradually gentler showers and longer sun breaks. Towering clouds turned dark at the bottom and emptied waterfalls, or hung droplets out for rainbows. When it was a freezing upper tier that cut loose, hail wiped out the paltry gains of spring. Along with snow that melted quickly, those outbursts were among nature's more boisterous declarations. They cleared the air of dust. Clouds were their marvels, collecting air-borne droplets, transporting and dropping them. Liquid versions of disintegration and renewal could turn quickly from vapor and droplets into downpours. What didn't stay local went into muddy rivers, and eventually to the sea. By mid-May, the sun went unchallenged for hours at a time, and, in small whirling pillars, the dust returned. Air arrived from the Pacific nearly dust-free, but soon stirred up more of it. By then, seedling firs had been planted, sagging fences propped up, and machinery overhauled. By mid-June, school-age crews were turned loose on strawberry fields in the first of the harvests.

A Mapped Wilderness

Once McLaurin had left the fields for the university, where buildup and breakdown cycles showed up in several study areas, he progressed by resisting new ideas and getting jostled forward as if boating on a choppy sea. He shed ideas of no further use to him, much as he did old work shoes. His better advances came in two quite different settings: the university and the wilderness. The first unaltered wilderness he encountered stood at the northern end of what had been a coastal Indian reservation set aside in 1855. Cape Lookout was its map name. Its nearly perpetual dampness was much as coastal tribes had known it for thousands of years. The rainforest canopy rose a hundred feet above an undergrowth of salmon berries, creeping leathery kinnickinnick, and much else he needed a field guide to identify - each growth, animal or vegetable following its own dictated cycle of growth and decay. Under mist rising from the sea, the turnover was heavy. In delivering rain measured in feet, the Pacific lifted clouds into the jet stream for transport inland, where the air at times felt damp and coastal for over a hundred miles.

Part way to the tip of the cape lay the scattered wreckage of a military plane the cape had snagged one fog-smothered day and was devouring. In the eerie quiet following the crash, vegetation started creeping into the remains. Planes could crash, ships break up on the rocks and sink, and people come and go, but the shoreline several hundred feet below would remain what it had been since the last ice age. The heavy growth and a population of slugs, worms, insects, and birds took command of small areas, home ground to them. The prevailing wind made the tall first declarative; for something incapable of getting annoyed, storming through and around obstacles, it sounded like troops of spirits protesting. The crumbling cliffs, the rhythm of days and nights, spring and fall, and trees three centuries old made time almost palpable. The cape itself went by rock ages as old as the sea. McLaurin didn't really understand the bending and blending of spacetime, but predicted that if he ever did get it down the Cape would figure in how he did.

Among the mountains and along that rugged coast, people could fish, hike, boat, water ski, and climb, but the sea and the mountains yielded little to them. On the cape's lofty perch, he felt on the edge of more than a cliff, as if of space itself and therefore, somehow, of time as well. For earthlings, being oriented meant knowing where the sun set, south pole to the left, north pole to the right, where west and east winds ran the sea and land exchanges, damp air one way, dusty air the other, storms coming in, rivers flowing out. *On edge* was how he felt where birds and sea creatures were at home and people were not. How could they be where vegetation and rust devoured aluminum propellers and iron engines?

CHAPTER SIX

WEST AFRICA, MEET EAST BRAZIL

The Argument

Naturalist overviews of terrain don't take it and its resources for granted, whereas economic and social histories and accounts of migrations and settlements normally do. To McLaurin, topography was too haphazard to support concepts of purpose and design. Earth history and early periods in anthropology both had random-seeming factors explicable in naturalist but not purposeful terms.

< >

The desire to see purpose everywhere is a natural one in an animal that lives surrounded by machines, works of art, tools and other designed artifacts; an animal, moreover, whose waking thoughts are dominated by its own personal goals. A car, a tin opener, a screwdriver and a pitchfork all legitimately warrant the "What is it for?" question. Our pagan forebears would have asked the same question about thunder, eclipses, rocks and streams. Today we pride ourselves on having shaken off such primitive animism. (Richard Dawkins, *River Out of Eden*, 96)

< >

When we say of a thing or an event that it was a chance happening, we do not mean that it was not determined by the laws of matter and force
We mean it was not the result of the human will, or of anything like it; it was not planned or designed by conscious intelligence. Chance in this sense plays a very large part in nature and in life.
The whole non-living world about us shows no purpose or forethought. (John Burroughs, in *This Incomperable Lande,* 237)

< >

Words and Things Again

Scholars in rhetoric, semiotics, and philosophy, concerned with how representations answer to reality, have sometimes turned to Emanuel Kant, Bertrand Russell, Martin Heidegger, Karl Jaspers, and Ludwig Wittgenstein and come away puzzled, especially from the last of these.

They would all have rejected what Richard Dawkins calls "primitive animism", which connects things human with things natural in invisible and mysterious ways. That clears away some of the odd ways of connecting human events to natural history, but still leaves those events and natural history leagues apart. Nor does it address problems in representation where natural history ranges beyond anything familiar.

Take the opening section of Wittgenstein's *Tractatus Logic-Philosophicus* (1999). It is cryptic enough to be claimed for quite different accounts of how concepts and means of representation hook up with real things: "The world is everything that is the case the totality of facts, not of things [...] for the totality of facts determines both what is the case, and also all that is not the case. The facts in logical space are the world" (29). Unless I am mistaken, this gets the initial premise backward - making the world facts and not things denies the priority of things over what we say about them. The empiricist assumption, in any case, is that what's out there comes first.

I'll take terrain as a case in point but, before we turn to it, let us first go further into how we nudge observations into facts when everything is working properly. Sensations precede recognition, if only by the brief interval from the beginning of a sensory episode to its reception. Assimilation and interpretation follow on the way to classifying and systematizing. How we ended up seeing and hearing so little needn't concern us, but it probably came about because the survival of our genetic ancestry depended on identifying certain things and not others. Hearing the whine of a mosquito came in handy, but hearing ultra-high frequencies and seeing gamma rays didn't. Indeed, as we've been seeing, nearly all the natural world lies outside perception and any conceivable pragmatic interest. For empiricism, the upshot is that although language indifferently states both facts and erroneous claims to them, the proof one way or the other lies with the phenomena. Intentional deceit differs from illogic and errors of perception, but joins them in lacking correspondence between what exists and what we think we see or what we say about it. By itself, general agreement about something means very little. At one time or another, people in nearly every quarter of the globe have thought that a particular sun god exists without anyone having ever seen one. Lack of evidence didn't prevent Re, Apollo, Huitzilopochtli, Nanauatzin, Jehovah, and many other sky, mountain, sea, and cosmic figures from being written into their respective cultural guide books.

Representations are ordinarily better organized than objects and scenes within the optical

range. That is the most noticeable difference between them. The following two images show a typical difference, one with a relatively high degree of regularity in an object, the other a computer-generated example of a highly symmetrical and intricate design:

A basaltic mountain spire is one thing - a constructed obelisk like the Washington monument quite another. Especially among living things, shapely objects aren't scarce, but generally speaking they are to geometry as a half moon is to a semi circle. Crystalline and frozen liquids can be exceptions, but nature's flat plains aren't quite flat and its spheres not perfectly spherical. Excess and random ebullience are other differences. For all practical purposes, life on earth could do without distant celestial bodies.

Some of nature's approximations to geometric shapes can be attributed to chance, on the grounds that if one looks at enough icebergs a roughly trapezoidal one might turn up. Without knowing every possibility, we can't be certain, but the chances of something highly proportional appearing among visible non-sentient objects are comparatively slight. An example of some spectacular beauty is Saturn's rings, but there again the rings up close, in images from spacecraft flybys, are ragged debris clusters, mostly of ice. Both disruptive and regular processes have put them there. That gravity and momentum arrange such miscellanies into roughly concentric circles is repeated on a larger scale in ring galaxies shaped around black holes. The principle of general approximation doesn't apply to underlying atomic matter invariables, however - they are rigorously governed by the strong nuclear force. The makeup of the elements, too, is regular on the level portrayed by the periodic table, and regular in the mechanical and deterministic sense that exactly the same thing happens every time water is heated to a certain temperature under a certain pressure. A highly disproportionate number of very exactly shaped configurations and rhythmic recurrences would imply that before they came about, they were on the drawing board of a master designer. In their respective cultural environments, Plato's Demiurge and the Hebrew Jehovah exercised that level of command, but, as we've been seeing, putting any such design behind natural phenomena raises problems with confusion, nature's robust harshness toward living things, and large areas of wreckage left by breakdown cycles.

As to Wittgenstein's apparent reversal of the priority of facts over phenomena, consider a typical sequence: after a long journey, light waves arrive, bounce off objects, and raise reactions in optic nerves. Only after the eyes and nerves have transmitted the reactions to the brain do we have what can be called an observation, which doesn't become a fact until it is confirmed. Such an observation can eventually get included in a proposition or theory, but most likely it won't be. To reach that level of formality, one observation has to line up with others, and together they must confirm what has made it through a previous screening process. Converted into statements, valid observations at last become facts and can be included in speculations, hypotheses, propositions, and theories. Receiving sensations without noticing them is more common than processing them that far. Peripheral sensations come and go every waking moment without our taking notice of them and sorting out reliable from doubtful ones.

Although everything large or small, fast or slow, illustrates that priority of atomic matter objects, topography has the most prominent collection of irregularities within our optical range. Land masses breaking apart and jamming together are medium size examples of buildup/break- down cycles that are much more massive and much older in celestial regions. The landscape of course constantly changes. That's no secret. Cliffs crumble. Earthquakes wrench two segments apart before the pieces once again settle. Erosion carves canyons and turns volcanic cones into jagged peaks. That much is clear without need of geophysics. Shoreline sands are disintegrated slopes, and stone is "sand and water, and a million years gone by," as the Beth Neil Chapman song has it. Something molten has to cool and harden before it can crumble into grit or move through the air as dust. Uneven cooling from earth's crust down to the core is responsible for volcanic plumes, tectonic plate movement, subduction, and rifts. The physics is always there of course, but we don't always need to consult it.

In short, terrain movement is discernible in its effects if not in its depths, or predictable in timing and magnitude. Except where badgers, moles, shovels, and mechanized land movers have been at work, earth's surface has been shaped by heat at work within and by erosion without. We wouldn't suspect from an initial impression that, over time, drifting land masses have made and unmade supercontinents from Vaalbara, Columbia, Kenorland, so on to Laurentia, Baltica, Rodinia, Pangaea, Gondwanaland. The evidence for that is substantial but not immediately obvious. No one knew of significant horizontal land movement until the 20th century, nor of convection currents and inner planet interactions of viscous and solid materials. The last feature to be recognized in geological history was the movement of solid upper plates down through and under softer layers by subduction. Volcanism goes deep as well, of course, and is likewise owed to uneven temperatures.

Before turning to how horizontal land movement became more understandable and could be integrated with evolution, however, let us sample traditional accounts of geography to fill out the steps needed to get from the phenomena to the continuum behind them. These set the typical stages that someone like McLaurin must negotiate.

Old Maps

Some of the discrepancies between actual terrain and how it was portrayed may have been artistic rather than intended seriously, but topography was nonetheless often claimed to have been allotted to a given nation by intent, a purposely vague one in the general semantic range of *destined*. Meant to be. The Babylonian clay tablet image of seven islands and a seven-pointed star in the *imago mundi* (about 600 BCE) overlaid geography with numerology as evidence of such a scheme. The second century (CE) Ptolemaic map extended the Mediterranean area east to west as a trade corridor, from the Iberian peninsula to India. It, too, made landscapes and seas out to look like fated dynastic territory. Europe, Asia, and Africa were placed around the sides of a lumpish square with Asia on the top. Ancient maps of Korea and China were similarly decorated with regal figures in command of sizeable territories, seemingly in league with undefined shaping forces. Astrology sometimes played a part in that, and of course the stars wouldn't be aligned on behalf of a ruler or a dynasty unless something very powerful was behind both. *Destined* was, again, stronger than *randomly* or *as it happened.*

Not until the post 1492 works of Abraham Ortelius (1527-1598), Gerardus Mercator (1512- 1594), and Hendrik Hondius (1573-1650) do we find the Americas well placed and not yet under a European colonial claim. They were soon to become the target of imperialism, carrying a missionary function along with regal and commercial ones. A king or queen could claim sovereignty over a sizeable remote land merely by financing a crew to sail there and plant a flag. For all that imposing of human-centered values on rock, land, and water, geography remained, as ever, just what geology and erosion dictated, and these in turn as inner thermodynamics dictated. Despite the flag planting and formal declarations of sovereignty that glorified land piracy, sailed seas in the colonial era remained unconquered. Climbed mountains remained the same mountains. The sun as an atomic matter body kept burning its fuel. A visitation by a ship or land expedition meant possession only in the minds of the voyagers and their sponsors. The dominion of the great white father in Washington that Lewis and Clark announced to indigenous tribes at each stage of their westward journey took many more years of migration, settlement, and military campaigns to bring about. Meanwhile, land controlled by the Sioux remained Sioux.

On Mercator's map (1569), South America remained a lumpish rumor, but the west African bulge on Ortelius's map (1570) matched the coasted shoreline of eastern South America. The Hondius map (1630), which portrayed the world in two circles sectioned by latitudes and longitudes, was as notable for imagination as it was for increased range and accuracy.

Surrounding 17th century maps were the usual regal figures and monsters of land, sea, and air, presuming impish or demonic influences. By 1630, sailing ships had coasted the west side of the Americas. Despite that, the overall rectangular shape of the Hondius map still reflected the four corners of the ancient flat earth, featuring its east/west division, further divided into northern and southern hemispheres. Farming, the arts, and the four elements appeared in conventional forms, and Oceania and eastern Asia were more or less in place, but mid-Pacific islands and sections of Australia were only approximated. Dominating an upper area of the Hondius map was a sun figure with a human face, paired with the moon as a lower counterpart. At each corner, portraits of monarchs were placed as if in co-rule with land-shaping forces. Once again, the implication was that terrain and water are governed not by natural laws alone but jointly by nature, humans, and higher powers. Imps and devils might or might not be involved in the misfortunes and the irregularities, but some of the topography and the storms at sea did seem devilish. The combined results were rugged terrain with hot and cold areas, jagged coast lines, and never-seen monsters of the land and the deep, all easier to put in storybook terms even now than in terms of heated composite buildup/breakdown cycles.

Despite the geology of Hutton, the Humboldt naturalist cosmos, and advancements in natural history in Lamark, Lyell, Wallace, and Darwin, some confusion continued in later 19th and earlier 20th century concepts of domain. Combinations of fact and fable produced the occasional oddity, like the later 19th century Professor Orlando Ferguson who cited 400 biblical passages condemning the "globe theory" or "flying earth," and offers in its place a stationary, square earth. For the price of a quarter in 1893, one could get his book and find how to knock "the globe theory clean out," a sentiment echoed even now by adamant flat earthers. A diagram resembling a map pictures that concept, unexpected at that late date considering it wouldn't have been before the circumnavigation of the globe and the mapping of continental and island shorelines. No matter how far they ventured, ships on the high seas never lost contact with their sovereigns (political systems) and gods (ecclesiastical systems), a practice built into the Patrick O'Brian fictional series along with a substantial naturalism, largely but not exclusively set in the South Pacific. As late as the early 20th century, misfortunes were still being considered providential in the narrow escapes of Arctic and Antarctic explorers. The assumption was that humans got themselves into dilemmas from which a watchful deity, appealed to in the right way, might rescue them. The two sides of ruling super powers, timely rescue and the unleashing of volcanos, plagues, diseases, blizzards, predation, and other rough and ready sides of nature, had to be kept apart. Topography and climate presented tangible and forceful obstacles to doing that logically. Deriving them from intent as always led to bizarre conclusions - without even putting aardvarks and protein eating sea urchins in the picture.

In taking topography for granted, mapping remains independent of geology and geophysics.

Its object is to show the results, not to sketch planet history. That the terrain has been taking shape for billions of years and is still in motion doesn't matter to a map maker or to migrations. As for the mapping itself, getting proportions and placements in accord with reality was a long time coming, considering the self-evident roughness of shorelines and mountains and the achievements by the late 1700s in nautical and land exploration. Nothing could yet be done to decipher micro mechanics, and the separating of continents couldn't be explained without the physics behind convection movement. Even the difficulties of making a sea-going clock weren't overcome until well into the 18th century, when John Harrison (1693-1776) invented a marine chronometer to help calculate longitude. That addressed a navigational and mapping problem independent of physics but not independent of the electromagnetic force in one of its manifestations. Where a compass gives only direction, shooting the sun, stars, and moon with a proper chronometer gives placement and relative motion. A sun thought to be orbiting the planet complicated the earlier calculations, but in any system a single straight line goes directly to a target, except for the undetected bending of space over long distances. Plain geometry always has at least that much to contribute to any form of mapping, geographic or celestial. With exploring, mapping, gathering of samples, and classifying of lifeforms, earth history and natural history were getting ever closer to the 20th century combining of astrophysics and geophysics.

That union couldn't be completed, however, as long as spirits remained aloft and in the seas, and the atomic matter continuum remained unknown. The celestial region had to be seen to have galaxies before the land and seas could be assigned their proper places in the forming of the planet.

In habitable regions, disregard of the prior rights of current land holders was the usual practice in colonialism, as it had been since the earliest consolidating of city states into territorial states and of territorial states into empires. Special backing from a high power wasn't absolutely necessary so long as some other disabling of prior human rights was available. Few, if any, of the expeditions finding a region already inhabited regarded themselves as the first pirates ashore, but the line between ethical and unethical possession was fuzzy at best. The biblical bunch of grapes episode (*Numbers* 13:23) shows the shock of that when the promised land turns out to be occupied by farmers and city dwellers. The genocide that follows (in the text, probably not in history) must have a high sanction indeed, and ancient Israel went as high as it was possible to go. Humans, as the planet's best adapters and most successful predators, had the organizing advantage of such beliefs with a similar anthropomorphic basis nearly everywhere.

Justifications for conquests still have contrasting views, inside and outside the invading country. Expanding nations have generally defended their aggression on principles made to sound higher than greed and ambition. Given agriculture's need to control the land, the enlarging of urban and farming domains was all but inevitable, but that needn't have come by militaristic means or involved genocide and slavery, practiced by some expanding nations and not by others or practiced at first and eventually dropped.

Strictly within the scope of national interests and without concern for how topography has come to be what it is, only a few recent earth-shaking events matter. Still, what has taken shape over time (to underscore the obvious) is what governs migrations and the establishing of territorial borders. On the west side of the Americas, mountains and deserts are paramount. A jagged northwest/southeast line extending for thousands of miles comes first and ways of life follow. It is complicated by transverse ranges like the San Gabriel and San Bernardino mountains of Southern California, but from British Columbia, south through Mexico and down the coastlines of Central and South America, movement inland encounters coastal mountains and higher ranges, the Cascades, the Sierra, and the Andes. Parched rain-shadow plateaus and inland basins make interior regions drier the farther south one goes in North America and part way into Central America before the tropical rain belt kicks in. Because climate and topography shape modes of production and culture, native tribes along the west-east line were less alike than those up and down the north/south coasts and valleys. A Kittatos tribe hunting deer and gathering berries in the lush rain belt of the northwest coast had less in common with Yakima and Palouse tribes and the equestrian Nez Perce than the inland tribes did with one another.

Continental Drift

That large earth plates move horizontally and subduct upon contact as governed by heat took a good deal to make convincing. In the earlier 17th century, mirroring shorelines were noticed as soon as they had been mapped, especially those of west Africa and eastern South America. In the next century, Hutton largely ignored those uncanny matchups and what might have been behind them. Darwin's predecessors were aware of topography as part of the species' governing environment, but it obviously wasn't up to them to propose continental drift. Humboldt speculated that interior pressure forced openings through the crust in fissures at the weakest points which accounted for mountains coming in chains, but subduction didn't appear among the uplifting and shifting forces for several more decades.

In the context of the continuum, the greater areas of celestial masses put earth movement in a different perspective. The Horsehead nebula contains an estimated 250 solar masses. The Orion is some 25 light years across and the Swan 15. The pillars of the Eagle extend 4 light years (or about 23.5 trillion miles). At 75 light years across, the Carina works out to some 436 trillion miles. These areas have massive internal movements as well as irregular outer shapes. Proposed black holes generate violent movements and add up to solar masses in the millions. As I mentioned earlier, the convergence of 14 galaxies occurred at about the billion-and-a-third year mark of the young cosmos. In that perspective, continents floating on a molten layer shrink in size. That wasn't broadly credited until well into the 20th century, each region having its own misconceptions of how terrain had taken the shape the maps by then showed. Without floating land masses, map readers from the seventeenth century onward, under the influence of one of the human-centered theories, found matching shorelines especially puzzling. The mirroring looked too extended to be accidental but too irregular to serve any imaginable purpose. That could be said of other topographic features as well, but the mapping of continental shorelines was more mystifying than most because jigsaw matches look crafted and deliberate [for more information on those who wrote about the matching African and South American shorelines, see Robert Wood's *The Dark Side of the Earth* (1985) and Hartman and Miller's *The History of Earth*]. Antonio Snider-Pelligrini, Osmond Fisher, F. B. Taylor, and Robert House, writing "On Earthquakes and Fossils" (1668), went on record as to what they thought. It seems that everyone had a theory, one of the oddest being that the dislodging of continents had come from the biblical flood. That wasn't inherently more incredible than continents floating on softened formations, but it required water somehow to work in under plains, mountains, and canyons and float what is heavier than it is. More promising contributions came from glaciation studies, vulcanologists, seismologists, and even from oceanography. To supplement what continental shapes and rock formations suggested, early 20th century zoologists and botanists charted animal and plant species on opposites sides of the southern Atlantic. Mineralogists compared rock formations. The evidence pointed to the separation of what had once been joined. No one was prepared as yet, however, to include entire continents forming, coming together, and separating in the buildup/breakdown cycles. At even just a casual glance, South America and Africa look as if a giant pair of hands had torn them apart. Any solid material might separate in about that kind of fracture pattern, but what could have done it? Uneven heat would be about the last thing anyone would have proposed.

Gathering data from topography and species didn't become a common practice until the later 19th century and (until Alfred Wegener (1880-1930) in 1912) not with enough thoroughness to propose a theory of plate tectonics, not so named or fully supported by geographers and geophysicists until 1967 by Dan McKenzie and Robert Parker, and 1968 by Jason Morgan. By then, horizontal land movement and the natural continuum had been recognized in many places for nearly four decades. With Wegener, drifting continents went from hypothetical to theoretical by means of combined disciplines, a major step toward acknowledging the continuum. What had become clear in Jean-Baptiste Lamarck and Charles Lyell – like Humboldt, influences on Darwin – was not only the dependence of species on environment but the dependence of one branch of earth and celestial science on others. To Lyell, the interdependence of disciplines was no different in the sciences than in historiography:

> An historian should, if possible, be at once profoundly acquainted with ethics, politics, jurisprudence, the military art, theology, in a word, with all branches of knowledge. . . It would be no less desirable that a geologists should be well versed in chemistry, natural philosophy, mineralogy, zoology, comparative anatomy, botany; in short, in every science related to organic and inorganic nature (6)

One problem with geology in that respect was that no one lived long enough to learn all the entwined disciplines, nor to witness any great number of revealing incidents. Among other factors, that was what prevented Lamarck and Lyell from anticipating Darwin more than they did, Lyell following M. Geoffroy St.-Hilaire (1772-1844) as an evolutionist connecting the "uninterrupted succession in the animal kingdom." The succession of plants and animals had, of course, adjusted incrementally to geography and climate, but their evolution wasn't yet a branch of the overall continuum. Lyell's discussion of species in the second volume of *Principles of Geology* (1830- 1833) came nearly half a century after Hutton had made the point that a credible theory of earth must draw on collaboration across disciplinary lines. Geology, biology, and astronomy in combination were stronger than any of them separately. Realizing that in reading Humboldt and Lyell aboard the Beagle encouraged Darwin to look closely at connections between species and environment in South America and the Galapagos. No one in the mid-19th century was prepared to see that both were products in some way of floating continents, but topographical shaping by volcanoes and earthquakes was clear enough. Hot spot geology was well ahead of a credible theory of lateral land movement, but neither sufficed without the other. Lateral land movement was as necessary to generate the earthquakes as vertical hot plumes were for the volcanoes.

Getting close to proposing horizontal movement was Humboldt himself, who, in looking at maps of the Atlantic, thought it resembled a valley. As a polymath, he came the closest anyone up to then had to linking planet history to cosmic history. Taking note of earth's inner heat, fissures, and vents, he could see that volcanoes bunched in clusters and lined up along what must be cracks in the crust. Adding electromagnetic force to heat, he explained the aurora borealis and, in doing so, made land and atmosphere interactive in a different way than climate and weather do by themselves. He lacked the necessary geophysics to propose a theory of plate tectonics, and the concept of a biosphere wasn't yet in circulation, but he went a step further than combining geography and geology. Bent on seeing whether all nature could be connected and aiming for a comprehensive theory based strictly on natural laws, – again, setting aside gods and demons – he entitled the last version of his work *Cosmos: A Sketch of the Physical Description of the Universe* (1858). He meant just that: "Nature considered 'rationally', that is to say, submitted to the process of thought, is a unity of diverse phenomena; a harmony blending together all created things, however dissimilar in form and attributes. The most important result of a rational inquiry into nature is, therefore, to establish the unity and harmony of this stupendous mass of force and matter" (24). Despite the emphasis on harmony, not a concept in keeping with nebular wreckage, predation, and natural perturbations like earthquakes and volcanoes, that was close to making *natural continuum* a term of terms something a steady-state creationist could concede. The mechanics were what interested Hutton and Humboldt and all that natural philosophy was prepared to accept. Calling in Vulcan or some other fire power improved the story value, but not the explanatory value. Even Humboldt couldn't quite stick to the phenomena themselves. In *Cosmos*, as in his other writings, he was often effusive and romantic in the manner of German literary figures of the times (Schiller, Fichte, and Goethe).

By 1912, the evidence for continental drift seemed sufficient to Wegener to propose it in earnest. In *The Origin of Continents and Oceans,* his matching of rock formations and of flora and fauna across the Atlantic in Africa and South America – a close enough match to indicate incremental changes in species over an extended period – supported what the shorelines suggested. The physics wasn't yet known, but what it needed to explain was becoming clear. Several more discoveries were required to bring geology and evolution into a more detailed partnership. Even so, the geodetic, geological, botanical, and biological data that Wegener assembled had enough non-coincidental particulars to be more convincing than its chief rival, the land bridge theory based on the raising and lowering of an Atlantic crossing. Rock that had extended that far would probably show the same differences on both ends that comparable distances do elsewhere. Neither hot spots nor land bridges had a realistic chance to account for detailed likenesses of terrain and species at a distance. Not yet mapped was a north-south mid-Atlantic ridge that plate movement created. Rifts, graben, and ridges aren't as revealing of what is going on deep down as volcanoes and earthquakes but, as age-old phenomena, their origins are the same. Thanks to explorations of the ocean floor, the mid-Atlantic ridge can now be found in encyclopedias like the McGraw Hill *Concise Encyclopedia of Earth Science.*

More explicit about continents once joined and now separated were Antonio Snider Pelligrini's contrasting maps, demonstrating before and after the horizontal separation of land masses. A single land mass consisting of Australia, Asia, Africa, and what he called Atlantide or the Americas looked feasible. In 1927, that supercontinent came to be called Pangaea and was surrounded by the ocean Panthalassa. In separated land masses, it was then called Gondwanaland (south) and Laurasia (north). The entire process, from the forming of the supercontinent to the current global map, took about 300 million years. Even that is only a small fraction of planet dynamics dating to over 4 billion years ago. Beginning with the first surviving maps, it took around 2700 years for geographers and geophysicists to outline a plausible sequence. Those two figures, the overall 4.54 ± 0.05 billion year age of the planet and a few thousand years of human puzzle solving, show the disparity between human and planetary events. Under empiricism, such gaps are filled stage by stage as a deterministic sequence. By the 1930s, the *geo* and *astro* subdivisions of physics had provided the means to do so. That fell short of determining what goes on in nebulas and molecular clouds, but it marked a significant advance in accounting for how solids, some rigid and some softened by heat, react to one another when vector and momentum bring them into contact. For that matter, a dropped vase, too, breaks exactly as it must, given the momentum, strike point, and lines of least resistance in the material. Under atomic matter invariables, what is true on earth would be true everywhere. Whatever the immediate concern, under the continuum *here* and *now* are also the long past and distant future, *there* and *then*. Earth, in that respect, becomes a geophysics laboratory.

Disbelief

Convincing as Wegener's case was in comparison to other explanations of topography, Gross and Harmon (2012) are probably justified in finding his presentation unsuccessful (although matters of that scope and complexity aren't entirely successful or unsuccessful). The reception involves sociology, psychology, and the dynamics of the academy. Adding to Wegener's case, the geophysicist Arthur Holmes published a study of relative viscosity that dispensed with the apparent contradiction of solid states plowing through other solid states.

Different levels of heat at different depths can muscle entire continents up, down, and sideways. When hard and soft masses converge, the effects can be as difficult to predict and account for as those of moving snow, slush, water, ice, and mud on a boulder-strewn slope of variable pitch.

Successive editions of Holmes' *Principles of Physical Geology* (1944) added to the initial proposal without charting all the necessary particulars of plate movement that figured in the 1967-68 proposals. The latter were new mostly in putting a few final pieces into place.

Any time after Holmes published his theory of relative viscosity, rudimentary geography could have included a reasonably sound theory of earth's age and land movement. The evidence cited by Wegener, Holmes, and du Toit in South Africa nonetheless wasn't fully credited by earth scientists in the US, Australia, and New Zealand until after midway through the century. Even the corresponding, widely separated rock formations of Africa and South America were considered to be only incidental – to a non-geologist, a debatable ranking of evidential priorities. The extent of a coincidence does count, or should. To Bacon in 1620, it already indicated that the Africa/South America shoreline match wasn't likely to be an accident. The problem lay partly in the fourth element of the object-perception-representation sequence, the reception (though largely in the complexity of the geology itself). As Gross and Harmon summarize, "despite Wegener's host of instances, geologists remained unconvinced because conformities didn't count, except as corroborative evidence" (Kindle edition, 3150). A certain amount of stridency in Wegener's style is understandable given that treatment of evidence. Conformities do, after all, make perfectly valid contributions, as the similarity of one puma to another is evidence of their genetic connection if not conclusive by itself. The conformity of one supernova to another is evidence of how mass and heat behave in an exceptionally large stellar body. The conformity of separated rock formations and their flora and fauna is evidence of a history spent in close proximity.

Even so, for reasons inherent in moving land masses, the resistance to plate tectonics wasn't as contrarian as it looks in retrospect. Without the right sequence and convincing geophysics, moving continents do look incredible, all the more so outside the context of much larger celestial mass movements. Breakage in the buildup/breakdown cycles comes in all sizes and, in nucleon decay, at every conceivable time span from millionths of a second to billions of years. As is now clear, land masses moving in what looks to be a random fashion are in keeping with the scrambled universe as a whole. The extended timeline and the dynamics of atomic matter were, nonetheless, too mind boggling to be immediately accepted. Even an adventurous Eve bringing war and pestilence to the planet through a deity's angry petulance seemed almost as likely as a drifting India plate pushing up the Himalayas. One might just as well think that a neglected rite had made Pele pour pahoehoe down on an offending village as that western California could wrench free of the mainland and cruise toward Alaska. That may not have been a problem with much of the scientific establishment by the 1970s, but it was at the popular level, which often settles for such humanoid waves of the hand as 'God's will' and 'God moves in mysterious ways', higher rhetorical ways of saying 'I haven't the slightest idea'.

Along that line, anthropomorphic versions of world history still remain more prominent than awareness of the long-term continuum. That isn't to say that all objections to plate tectonic details have been unjustified or based on an anthropomorphic principle. With respect to plate tectonic theory in general, John McPhee (1998) cites Anita and Leonard Harris, both critical of geologists who apply it to every planetary wrinkle, especially questionable with North America's east side. "I believe in plate tectonics," Anita Harris concludes, "just not in the way they're perpetrating it for places like the East Coast. It shouldn't be used as the immediate answer to every problem" (231). Nonetheless, no combination of factors works as well for both vertical and horizontal land movement as rising plumes, eruptive vents, earthquakes, and subducting plates. Spin the time clock through the ages and watch earth's crust demonstrate that, or turn to Jupiter's too-close moon Io, squeezed by the huge planet's gravity into a heat that raises storms of volcanoes. Apply that atomic matter heat-governed turmoil to earth's earlier history, now cooled to a quieter phase.

Despite the complexity of geologic history, billions of years of wear and tear on the planet can be summarized in a brief paragraph, done nicely by Reg Saner in *The Four-Cornered Falcon* (1993): "Like everything born of fire, our planet is cooling. Eventually its 'secular cooling' will bring continental drift to a standstill. As convected heat within Earth's mantle and outer core loses its radioactive enthusiasm, the lithosphere under the sea and land will thicken. Once continental plates lock in place, volcanoes won't erupt nor will new mountains thrust up to replace old mountains worn down. Thus, our planet's flat future; its alps will be molehills low as a hundred feet or so above sea level" (157). Sand, grit and gravity will triumph over high peaks if not necessarily over canyons and gullies, some of them so deep and renewable by runoff that filling in could take even longer than carving out. A few lumps will recall the mountains, and a few grooves with rounded edges the canyons. We need multiplication with a high exponent to get that far into the future. The sums needn't be precise to suggest that it will be a long time indeed before canyons fill in and mountains lie down flat. The governing process, as always, is the alternation of heat-powered buildup and breakdown from the outset of the universe to its quiescent end.

We've only begun to assess the implications of shifting land masses for social orders and economies. I've been lining up old maps and the history of horizontal land movement partly for that purpose, to enable the natural continuum and planet history to have their impact on migrations and on the buildup and collapse of civilizations. In the extended story, the placement of islands and continents directed animal movement and set living conditions. The long view isn't necessary to every purpose of course, but it is for getting the scalar proportions right and natural philosophy squared away with the continuum.

The Continents' Current Locations

Not only do objects come before sensations and facts, but as the foregoing indicates, geography comes before economics, long before as in economic terms something advancing in mode of production intercepts something traveling in geologic time. Consider the intersecting and overlaying of spans in almost any encounter of social and economic factors with geographic ones. Multiple courses converge on a wagon train headed west in the 1840s, flora and fauna with their individual species evolution histories, formerly Asian native people, recent immigrants from across the Atlantic, and inventions, especially of weapons and iron tools. Or, consider an explorer crossing a crevasse on an ice bridge. Like everyone else, he is following a personal lifetime clock, but in that case amid an older depth-of-ice time chart, possibly thinking of fame and fortune good for a few decades. What we set as a time span produces the pertinent biographical and economic factors, taking the long term for granted. The deeper the time, the more generic the results. At depths beyond anthropology and archaeology, at ages written in stone, reconstruction is turned over to geophysics, mineralogy, and other earth sciences.

Hence the reminder: phased and stacked multiple chronologies intersect in everything we do without our paying attention to them. Nothing methodical can handle the crisscrossing lines any more than measurements of gravity can include all gravitational masses within range. We take the long range for granted. That's as it must be, since it isn't the function of historical chronicles to go beyond immediate motives and conditions. These are abundant and tangled enough on their own without our considering the Grand Canyon. In long span terms, the distance from Europe to the Americas delayed the crossing of the Atlantic until sailing ship technology and armed crews with muskets had monarchies behind their expeditions - a history of European migrations can justifiably ignore that. By discipline in planetary studies, rock-based science fills in the million-year range, chronicle histories a moderate range of centuries and decades, diaries and anecdotes the personal day-to-day range. The number of factors that intersect in any given expedition is incalculable when we include economy, politics, technology, and intersecting biographies. With so much on record, the more prominent factors concerning geography, human history, political orders, technology, and economics are readily available - many of them retrievable electronically in a few seconds. Again, just to be clear, we don't need them for most purposes. Natural philosophy, however, is guided by the relative measurements of the continuum and the scalar spectrum as a whole and therefore acknowledges the framework and selects for citation whatever a given purpose calls for, confident of the succession.

Increasing the length of a history shrinks the given event; this in keeping with objectivity, which in its comprehensive dimensions discards individual and national achievements. Science and philosophy are indifferent to individual credits, even if scientists and philosophers aren't. In the context of combined historical and geophysical elements, it is clear that, beginning with the 17th century, a combination of culture, technology, and geographical location put North American in prime position for colonizing and eventually world trade.

Cultural and economic development, ancient in Europe and Asia, joined with geography in making the Americas ripe for exploitation, which (among other things) meant displacing hunting and gathering tribes with builders of cities, manufacturers, and international traders. With that prying open of the New World, age-old imperialist ambition found a new venue. Competition among monarchies was an additional factor, but among oligarchies and democracies new world development might have gone much the same. In continents that had fortunately turned out not to be Asia, traders found markets, and immigrants found places to settle. People who needed masts for ships and firewood found trees. Those who farmed found land. Those who wanted to strike it rich found gold, silver, and copper. What unfolded as individual histories were also geographic, technological, national, and epochal histories. The natives had to be moved aside and that often got violent, but something like *Manifest Destiny* was always there to sloganize the operation. It was indeed destiny, in the sense of many things converging in a given era within an overall geographic logic (the latter based on the separation of the Americas from Asia, Africa, and Europe). Lettered societies with maps and books have a decided communicative and organizational advantage over people lacking written documents.

In brief, the west-facing shoreline of Europe and the east and west facing shorelines of North America were well placed for a certain stage of technology that featured ships, trains, and immigrants armed with muskets and rifles. The interior was ripe for settling by those versed in farming, mining, industry, and who, in contrast to roaming tribes, had national administrations.

The US and Canada, closest to northern population centers in the Far East and Europe, became global crossroads, their prosperity thereby not guaranteed but enabled by the geography. Where New World timber, metals, and crops provided the raw materials, Asia, Europe, and Africa supplied some of the technology, the manpower, and the markets. The availability of the Americas for that late exploitation was due to their being beyond the range of earlier migrations out of Africa that had populated Asia, Oceania, and Europe. Weapons, sails on masts, a venturesome spirit, ingenuity, courage, and ruthlessness did the rest.

In contrast (to place a quite different case within that frame of reference covering several centuries), the imperialist ambitions of the Third Reich came at a technological stage of manufacturing, newsreels, aircraft, communications, and shipping that allowed a concerted effort by the Allies, east and west, to mount a counter offensive. What the leading Third Reich predators thought to be looking ahead was actually a re-enactment of land piracy, several thousand years old and not as likely to succeed in Europe in the mid twentieth century, despite a head start in armed forces and fascist fanaticism coordinating the effort. Like several other imperialist movements, this one too had a holy crusade element, featuring fear and hatred and fueled by economic depression and an unpopular World War I settlement. Jews had the wrong religious doctrine and Marxists the wrong ideology. Against that fervor, allied troops and arms from the east and from North America were likely to be heavy and effective. Technology hadn't yet produced intercontinental missiles and nuclear warheads – if it had, the devastation would have been even worse. As it was, a war of empires, carried over from World War I, brought more widespread misery than any previous conflict had. It was within the Nazis' and fascists' power to spread havoc and bring unparalleled suffering, but not to establish an Aryan-dominated world or even a lasting European empire. That was, again, an intersection of differently paced movements, continents, nations, science, technology, and an us/them psychology underway almost as anciently as drifting continents.

At the University

In the last chapter, McLaurin's introduction to the universe marked a turning point, when he realized somewhat haphazardly that what lay behind everything extended indefinitely into the past. The backward trail of sources went forever - not that he needed to take it. Nothing hung on his ignoring it, but philosophically that made a difference. He couldn't help noticing how much nearly everyone he knew depended on some long-range belief that stood little chance of being true. The pieties sounded somewhat hollow; they also seemed to be unnecessary. Even those who swore by them tended to set them aside and proceeded with the business of the day. It was only when they paused and thought big that the beliefs returned, as if pulled down from a shelf where they had been stored and gotten dusty.

As an undergraduate, he too wondered if the works that artists, carpenters, engineers, and builders of any kind, for that matter (including termites building nests and bees building beehives) produced might not be matched, if only roughly, by emergent designs in nature. After all, stars gathered into spheres from whirling accretion disks and had satellites moving around them in more or less regular orbits. Galaxies formed into vaguely geometric swirls. Trees at the insistence of gravity came in balanced shapes as did quadrupeds and bipeds. So much symmetry, so much logic, so much beauty and regularity. Generations in every species more or less repeated one another on schedule. The spirals and decorative patterns of sea shells fit such notions as did maple leaves and snake skins, the latter intricately patterned. Indeed, it wasn't hard to compile a long list of well shaped objects. Snowflakes and crystalized minerals carried design into the intricacy of non-sentient objects. At the building block and chemical levels, composites were strictly ruled, seemingly without fail. If domesticated plants and animals could be bred to order, and if mathematical configurations were more than chance accounted for, might not natural selection have been designed to move ever upward? How could all that regularity and symmetry come about without some sort of blueprint design? Sizeable segments of the population seemed to think that had been the case throughout a guided, purposeful evolution, reaching its current pinnacle in mankind. Historically, some reputable philosophers, poets, and theologians had also assumed that to be the case – in fact, most of them.

He eventually learned, however, that thinking of that kind had been debunked by Thomas Huxley and others in the later 19th century. With or without a master plan of some kind, some school districts in the US still had reservations concerning the teaching of evolution until very recently. In any case, it didn't look to be going anywhere not explained by environment. No cabinets grew in forests, nor, despite his admiration of Robert Browning, Malthus, Matthew Arnold, and other Victorians, could he find trials and tribulations looking purposefully staged to test and strengthen character. Plagues, pestilence, wars, drought, and the general run of diseases and accidents were too severe for that, and ailments in flora and fauna could scarcely be called character building. Collapsing all the higher powers and divinities that people had adopted over the centuries into one being forced that one to be multifaceted enough to account for the entire range of nature. One occasional alternative, especially in earlier times, was to combine a purposeful creation or directed evolution with demons canny enough to do their work without anyone catching sight of them. That seemed, to him, too far-fetched to be given a second thought.

It wasn't in anything scientific or philosophical, however, that McLaurin entertained the possibility that natural and human history might fall into configurations more often than accident and natural laws accounted for. That came in a Shakespeare course of all things, and went like this: in sophomoric fashion he came to realize that, despite mirroring a messy and often unjust world, the plays fit their scenes together meaningfully and came to decisive endings. Hamlet did and said very little of anything unrelated to the task at hand or to his delay in performing it. Allowing for theatricality, the richness of the language moved an ensemble of details forward in purposeful increments with a realistic look. King Lear, too, worked through alienation from daughters one, two, and three in fairy tale fashion, and contritely back to the good daughter with an inevitability as convincing as it was gut-wrenching. Again, except for the theatricality, that looked and sounded so real it could possibly be mirroring reality, chosen and manipulated to forward a plot but, at the same time, a reflection of life – meaningful, though dismal.

Perhaps finding the right vantage point would reveal something like Causabon's key to all mythologies (in George Eliot's *Middlemarch*). In Sunday school at a rural hillside chapel's Sunday school, the few times he had attended as a pre-teen, something like that had been assumed if only in a vaguely pious way. Members of the congregation presumably carried it home to ponder, or perhaps merely set it aside until next week, same time, same place. How that hour applied to life was up to the individual, but it looked possible in literature for *mimesis* to be *extracting* patterns *from*, rather than *imposing* them *on*, its materials. That notion had the backing of Aristotle as well as some reputable literary theory. Have faith: even misfortune and seemingly random events will eventually be explained. Dante and Milton, encountered in his junior year, imagined a purpose that had been obscure coming to just such an eventual clarification. The end would not only clear up mysteries, but justify what had occurred along the way. Well into the 20th century, Pierre Teilhard de Chardin and Pierre Lecomte du Noüy endorsed variants of evolution emerging purposefully.

Could there be *nothing* to such notions?

First, the coherence of literary texts wasn't nature, but art;tIt had realistic elements, but was the doing of writers rather than cosmic forces. Nothing comparable appeared in the wild. Keeping an open mind was a student alternative for those tending to posing big questions in late night dormitory discussions, and meaningful coherence did, after all, have a will-o-the-wisp quality - appearing, vanishing, reappearing. Nothing, however, succeeded in explaining or removing large areas of chaos and catastrophic episodes or, for that matter, routine breakage and suffering. Even if geography's irregularities were somehow explained away, as centuries of ingenious commentary illustrated, when one looked to the heavens the Magellanic Cloud and Cassiopeia added to the problem by removing mankind and any concept of morality and justice from the picture. Unrelated to that, but as an interesting exponent of traceable order and its limits, chaos theorists had recently gone to work. Atoms were more rule-bound than larger composites, but thanks to quantum physics and indeterminacy (1927), even that level of the scalar spectrum had problems. Whatever conclusions natural history eventually reached, it would always seem strange that an electron could be both a particle and a wave. No one could say for sure what it was because the act of observation turned it into something else. People, cats, and dogs, in being observed, put on a certain look - but should a mere electron be doing that?

Despite the generally temperate climate and spectacularly rugged mountains and coastline, it was as clear in the Pacific Northwest as anywhere that geography was random in appearance and the biosphere was deadly. A few miles into the foothills where McLaurin grew up were pre-agricultural and pre-deforestation rain forests. Symbiotic cooperation in those areas came mixed with competition. By driving a few miles north, students could sample mercantile and industrial Portland with the usual civic order, street grids, daily scramble, crime, and topographic irregularity. A few miles southeast brought them to mile upon mile of broken lava fields. A hundred miles east carried them into arid deserts with ravines, rimrock escarpments, and dry stream beds strewn with boulders powered there by floods, none across the West larger than the Missoula floods. Long ago, these had carried soil and debris all the way from Montana to the Willamette Valley, sweeping life before it in the process. In tide pools a few miles west were archaic forms of life hiding in cracks and crevices, seeking to snare prey or to avoid being snared. Thousands of years of wave action had hollowed out rock formations, with results nearly everyone considered beautiful but obviously haphazard. Though patterned butterfly wings supported equating truth and beauty to some extent, that and other versions of symmetry in living things came limited in scale and had evolutionary histories that explained them on harsh Darwinian principles.

Even some who lived near the volcanic peaks of the Pacific Northwest didn't fully appreciate the combination of unleashed natural force, spectacular vistas, and disruption – that is, until the 1980 eruption of Mt St Helens blew the top off the mountain and left a caldera similar to one Mt Mazama had left at Crater Lake. What showed in that spectacle, as in other events of magnitude, was as fundamental as eruptions in stars. Solar flares, X-ray and gamma-ray emissions, white dwarf implosions, and storm-stirred nebulas were featured in an astronomy course he took to fulfill the breadth requirement. They had been the doing of nuclear fission and fusion at prescribed heat and mass, a thermonuclear principle that prevailed everywhere. Seen from within a hundred miles, the exploding mountain looked uncomfortably close, as if it belonged to a time when creatures half bird and half dinosaur roamed the earth. Volcanic ash drifted onto rooftops hundreds of miles away. Powerful earthquakes in San Francisco in 1906, Long Beach-Compton in 1933, and Alaska in 1964 had also reminded grounded creatures that the earth beneath their feet could suddenly become treacherous. Those temblors and the Loma Prieta quake in the fall of 1989 attracted more than local attention, but the general rule for salience and proximity held; similar catastrophes far back in time or halfway around the world made comparatively little impression. Subjectivity adopted that as one of its sacred commandments. Equivalent disturbances on other planets and in other galaxies had little effect, except on a few devoted to astrophysics.

Disruptive events that were only predicted and not yet historical had a mixed effect as well, like the one in the offing along the Hayward fault through the dense populations of Oakland and other east bay cities. That was more to the point when he looked forward. None of those things were over and done with. He would be running into them off and on for the rest of his life.

Potential disasters were nothing like one in progress, but in natural philosophy, along with past events that had left their mark, all such weighed the same. That interplay of the core self and atomic matter was complicated by something far off drawing near quickly. Along with three fourths of other species, the dinosaurs were most likely wiped out in short order by a single catastrophe at the Cretaceous–Tertiary (K-T) border. No thorough account of what the universe contained could overlook such intersections - or, for that matter, the chicken that chose the wrong moment to cross the road. The principle of safe distance worked on temporal as well as on spatial terms. Earth's greatest disruptions had come early in its history. Their gradual winding down was the result of cooling and of earth's orbit getting swept reasonably clean of larger chunks of debris. Visible features of landscapes, such as Burroughs (another later discovery), pointed out were random in placement and shape, certainly not planned by something using a geometric template. That more land lies "in the northern hemisphere than in the southern, more land at the South Pole than at the North, is a matter of chance." So was the overall "distribution of land and water, of islands and continents, of rivers, lakes, seas, mountains and plains, valleys and hills, the shapes of the continents" (*This Incomperable Lande*, 237). Chance and destiny combined when the planet, on a certain vector, collided with chunks of star debris as large as mountains.

Fortuitous intersections showed up as far through the universe and back in time as telescopes reached.

Humans were an adaptable, clan-forming, group-believing species. In the first prolonged human migratory phase, clans had followed surface wrinkles and water courses in locating hunting grounds with favorable geography and climate. In the migratory/colonial phase, from the 16th through the 19th centuries, explorers mapped courses and measured distances with the aim of locating trade routes as well as seizing occupied land, carrying their beliefs along with them. For several centuries, explorers had been encountering topographic irregularities without placing them in long-range geophysics, or giving even a passing thought to multi leveled encounters in which breakdown at one level contributed to buildup at another. The harrowing stories of John Franklin, Robert Scott, Apsley Cherry-Gerard, Douglas Mawson, Ernest Shackleton, and the George De Long's expedition were prominent among those who sailed, walked, and sledded. They, too, sometimes found what they thought might be shaping influences in landscapes and natural events - despite frozen regions of broken ice, crevasses, seracs, and sastrugi. With or without some sort of high-powered oversight, inhospitable icescapes put miles in perspective. Douglas Mawson pulling himself out of a crevasse up a knotted rope hit, what felt like, the utmost limit of human endurance, only to drop back down and have it all to do over again.

European explorers were astonished and sometimes dismayed by what they saw. Graham Collier found Peter I Island forbidding in the extreme:

> If time is measured here, it is in millennia, distance in light years. The two small, insignificant ships, hundreds of miles from land in any direction, even further from anything we might call human, could easily become little more than matchwood if the pack ice began to move.

> It is difficult to explain the sense of isolation which comes from being so far from anything familiar–facing an environment so unforgivingly bleak and hostile. Everything here, the sheer cliffs and volcano in their blackness and redness, seem to be inimical to everything we have learned from Western civilization. Religion, music, art, science, gentleness, kindness, compassion–all are rendered meaningless in the face of this harsh and utter loneliness (*Antarctic Odyssey* 49-50).

That sense of alienation may have stemmed from assumptions of "religion, music, art, science, gentleness, kindness, compassion," but it was hard for any non-swimming creature, civilized or not, to welcome much of anything within the Arctic and Antarctic circles. Creatures suited to the sea's terrain and climate could survive there, but not comfortably, and being suited for a polar region made them less suited for temperate and tropical regions. Since nature displays similar features in many places, another set of assumptions might have been confirmed rather than violated by Peter I Island, which was, in any case, more hospitable to life than any known place off the planet. Once explorers had been exposed to frozen wastes and jungles, and Darwin had established natural selection as a cruel process of elimination, thinking of nature as red in tooth and claw became common. That too was selective. It took minimal experience and no science to see that the proportions of habitable and hostile terrain, order and disorder, competition and cooperation, pattern and chaos, varied a good deal from one setting to another.

McLaurin was impressed by the beauty of some of the irregularity, but never again tempted to find natural wonders purposeful. *Harsh beauty* wasn't an oxymoron. He found fractures, twists, and bent and distorted geometry of more interest than squares and circles. The misshapen and the knotted was harder to put into words or numbers, but he knew of those who could do a certain amount of that. Irregularities were figuring more and more in applied mathematical calculations and in diagrams, especially those of chaos theory. Weather forecasting had worked its way into how progressions were calculated in other fields, as in the strange attractor graphs of Edward Lorenz depicting what was both deterministic and unpredictable. When a colleague who professed sympathy for French linguistics and deconstruction lamented that no representation could be trusted, that reality was one thing, the mind and its symbols quite another, McLaurin countered that going wholesale in distrust crippled communication. It was true, however, that much that came to hand seemed incompatible with what major branches of intellectual history had been assuming all along.

Except for a few creatures with architectural talent like beavers, birds, bees, and people, natural history couldn't have produced anything other than its mixtures of order and confusion. Terrain came with canyons, gentle slopes, level plains, and rocky crags. He admired those who had pioneered courses through it and became more appreciative of catalogue systems, tables of contents, and indexes–small weeded patches. They had nothing like beauty, but they performed a necessary organizational function. He had looked forward to lock-down knowledge in theorems, precepts, and numbered and chaptered histories. Roughhewn nature was teaching him to settle for less.

CHAPTER SEVEN

PATTERNS FORMED AND BROKEN

The Argument

Non sentient patterns are strictly mechanical, the living ones combinations of biological or botanical necessity, habit, environment, and, in some animals, of individual and group choice. As examples in war and strata and as bifurcation charts indicate, once division and multiplication advanced into intelligent life, disrupted, inverted, and complex patterns multiplied beyond accountability except in categories and lumpsum estimates.

< >

Nature forms patterns. Some are orderly in space but disorderly in time, others orderly in time but disorderly in space. Some patterns are fractal, exhibiting structures self-similar in scale. Others give rise to steady states or oscillating ones. Pattern formation has become a branch of physics and of materials science, allowing scientists to model the aggregation of particles into clusters, the fractured spread of electrical discharges, and the growth of crystals in ice and metal alloys. (James Gleick, *Chaos*, 308-309.)

< >

In Summer and in Winter I shall walk Up and down
The patterned garden paths
In my stiff, brocaded gown. . . .
The softness of my body will be guarded from embrace By each button, hook, and lace.
For the man who should loose me is dead, Fighting with the Duke in Flanders
In a pattern called a war. Christ! What are patterns for?
(Amy Lowell, "Patterns")

< >

Buildup/Breakdown Cycles and the Pattern Called War

For recurrence, generally *scheme, shape,* and *configuration* would serve as well as *pattern* so long as we remember to include temporal as well as spatial regularity. One advantage of general terms of that kind is that, besides applying to all scalar levels, they work for both organic and non-organic things - and equally well for temporal and spatial configurations. Familiar visual patterns like strata are simple and ubiquitous, as are regularities like pulsations, oscillations, and orbiting cycles. War makes a striking example, in that it is disruptive but paradoxically *ordered,* creating its havoc *methodically.* It matches other behavioral and non-sentient patterns in that regard. Throughout the universe, wreckage is rule bound and thus resembles warfare – except, of course, not intentional. The buildup/breakdown cycles are evident everywhere except in a few things that never break apart, such as unattached neutrinos and hydrogen. When supernovas disintegrate, they provide the debris, dust, and gas that gather into second generation stars. At the opposite end of the scalar spectrum, atoms of a lower atomic weight are disassembled to build heavier ones. Proto stellar and proto planetary disks are in-between states. The disassembled stuff of former stars collapses into denser areas of dust, flattens into whirling disks, and gathers into gravity-shaped spheres, leaving unassembled areas of debris. Spectacular results can be seen in photo galleries online, as in the "Pillars of Creation" and "Orion." In general, the continuum has produced a prodigious multiplication of things, as evident in the number of galaxies, suns, planets, and moons plus left over debris, and in the evolutionary tree, beginning simple and producing species in the millions.

Until chaos theory got seriously underway in the 1970s, science and philosophy lacked an organized way to translate complicated patterns into diagram format. The essay collection *Dynamical Chaos* (1987) illustrates how far into irregularity the chaos theory methods and charts go. Steven Strogatz, in *Sync: How Order Emerges from Chaos in the Universe, Nature, and Daily Life* (2003), speaks of "synchronized chaos" and maintains that "chaos *can* sync" (183). Indeed, it can in limited ways, short of utter chaos. In terms of the scalar spectrum, another general principle that combines confusion and order is that one level can fall into disarray even as another remains regulated or grows. Thus, agitated molecules, besides being individually regular, assume the shape of whatever contains them. Two pattern-*challenging* factors to keep in mind are collateral damage and unanticipated intrusions from outside an enclosed system. The latter is illustrated by an object colliding with something unrelated to it. Concerning the former, the harnessing of atomic energy to blow up enemy industrial complexes was intentional in the 1940s, but the targets didn't include commercial and residential areas that were likewise destroyed. Melting ice and rising seas are a byproduct of human activity that isn't intended to warm the planet. The spread of communicable diseases wasn't the reason cities were built. This works in buildup as well; light and heat cast off by the sun wasn't intended to feed into green growth.

Both non-sentient phenomena and lifeforms constantly reconfigure elements in this way. The soil where much of the planet's breakdown phase ends up is a storage compound, disorganized in itself but available for multiple uses. Its materials are in an interval condition at the moment, neither gathering nor disintegrating. Whether or not something broken loose from a former composite ends up in another depends on the circumstances and on what the new buildup calls for. The next phase of discarded materials isn't built into the separation or the breakdown, but into what a succeeding configuration calls for. Where the milk goes that leaves the cow isn't determined by the cow.

That has an odd subset in how living things range abroad for what they need. Scavenging is among the order-creating forces, this one not as well recognized as it warrants. Non predator scavenging, such as grazers cropping grass, is like someone visiting a wrecking yard for auto parts or a historian thumbing through materials from multiple sources to assemble a refurbished history. Wandering opportunism like that can be as basic as something in motion severing its connection to one location in moving toward another. Changes in location reconfigure objects in the altering fields, as what is a tangle of branches, limbs, and leaves in the midst becomes a shapely tree from farther off. The new perspective doesn't collect parts from here and there as scavenging does, but by changing the distances and angles of vision it assembles the parts in different configurations. On the spot, placement in that respect is pattern making. On the move, it is simultaneously pattern making and pattern breaking.

One brief reminder of the bottom level: patterns of all kinds (placid, turbulent, peaceful, warlike) get underway at the building block level – as, of course, everything does. Not only does composite buildup start with quarks in protons and neutrons, but disintegration ends there. What physicists mean by symmetry at that level isn't the same as spatial geometry's symmetry, but is nonetheless a fundamental form of regularity. Even anti-patterns that upend a given symmetry follow cycles, as Dave Goldberg's inventory of things inside out and upside down demonstrates in *The Universe in the Rearview Mirror* (2013) and *A User's Guide to the Universe* (2010). A good deal in atomic matter has an opposite, as the antineutrino and the antielectron (positron) stand across from the neutrino and electron, without in those cases warring with them, merely opposing and locking in a structure of paired sets. Still more broadly, Lederman and Hill (2004) praise the "myriad incarnations" of "innumerable patterns designed by nature" (13). Under those conditions, symmetry is "an invariance of an object or system to a transformation," *invariance* meaning that the state of an object remains the same after a transformation (15), as a symmetrical mass rights itself like a buoy that keeps returning upright despite being buffeted by waves.

Symmetry at the atomic level receives a serious qualification, however, when Lederman and Hill (2013) turn to sizeable objects in the "real world of planets, stars, iPhones, and humans . . . a world of broken-down symmetries, as though we live among the ruins of some ancient civilization" (99). Only "in the first instants of creation" when "all particles were massless" were the symmetries intact in that sense (100). The pre hydrogen stage has also been characterized as a hot plasma fury, symmetrical only in that all the particles were the same and moved as they had too. When Leonard Susskind (2006) reviews the heyday of atom smashing and what it discovered, he finds the particles "an unruly cast including electrons, neutrinos, muons, tau particles up-quarks, down-quarks, strange-quarks, charmed-quarks, bottom-quarks, top-quarks, photons, gluons, W- and Z-bosons, Higgs bosons, and a good deal more. Don't ever believe anyone who tells you that elementary-particle physics is elegant." This hodgepodge "reflects an equally unwieldy jumble of masses, electric charges, spins, and other properties" (50). The particles are uniform only within each kind. In combinations and in what they build, they are well on the way to nature's object-size proliferation. Still, the basic rules of the assembling never fail.

Rules and the Unruly

A pattern obviously doesn't have to be exactly regular to avoid chaos. Whatever its tempo, the oscillation of a flag flapping in a wind holds well enough to follow visual and aural patterns. Even a loose fragment of something on the fly isn't chaotic by itself, since it has a computable velocity and the geometry of its vector. Whether or not it is part of a chaotic field depends on the rest of the field. Even a cloud of dust isn't completely chaotic, since its specks move more or less in unison and are similar in size and makeup. Swirls within it are sub patterns below the confusion level that hundreds of them in combination would create. Flow patterns governed by density and gravity and not too disturbed by collisions are reasonably orderly but also irregular and disturbed. Adrian Bejan's and J. Peder Zane's constructal law (2012) maintains that "Everything that moves, whether animate or inanimate, is a flow system. All flow systems generate shape and structure in time in order to facilitate this movement" (3). They follow universal laws in continuing motion until obstructed and in taking the path of least resistance propelled by gravity. In conjunction with obstructions, slopes and fluidity determine their velocity. From the perspective of something finding its way downhill, the resistance is partly random, as a mountain stream fortuitously tumbles over and around boulders. Being gravity ordered, the fall is regulated and the stream channel is more or less stable, staying in the same course at roughly the same velocity, but the actual movement is rough and tumble. Any recurrence, even of that irregular kind, constitutes a pattern. In a delta, the movement of water takes several different courses coordinated in direction and reaching the same spill-out. We grow accustomed to such mixtures of confusion and pattern and scarcely notice them once they have been explored and mapped. Much of life follows similar mixtures of regulated pattern and disturbance.

In that respect, an excellent question, Amy Lowell's what are patterns for, and "pattern called a war" qualifies as one of Burke's perspectives by incongruity. War's well-drilled armies and designed weapons disrupt targeted patterns of many kinds: architectural, political, military, biological. The better the stratagems and more disciplined their agents the greater the havoc.

When bombers fly in tight formations, the pattern above methodically destroys patterns below. The conduct of civilized war follows protocols set down in articles of war. In boxing, footwork, jabs, and uppercuts follow patterns designed to deregulate an opponent sufficiently to convert vertical posture into horizontal. The end of hostilities in a war is as ceremonial as their beginning. Ceremonies and speeches attend the burials of the maimed and destroyed bodies. Victory in a boxing match is signified by a ceremonial raising of the victor's hand. In war, monuments in stone and rows of crosses enshrine the names. In boxing and other sports, ranking organizes the participants. The commemorations and memorials in any context reinforce the motives for the next round of enlistees. They, too, will be praised, as in warfare they go in coordinated ranks into a melee to encounter those organized in a similar fashion. In general, collisions based on vector and velocity range from atoms and molecules to merging galaxies. The disruptions, too, often follow rules and regulations.

Whether or not war making and forms of competition can be traced to something built into the human DNA is neither a hopeless nor an entirely answerable question. Among the more puzzling combinations of pattern making and disruption, I've singled out war partly because, besides using method to inflict ruin, it connects recent group behavior to ancient team hunts and the ubiquitous I/you, us/them polarity. That is based on the placement invariable, the location of each and every composite - hence, in living things, the territorial instinct. Teamwork is generated by the proximity of tribal members, given at times to skirmish against neighboring tribes. The self is an integral, living version of that placement. Even without teaming up, it can be defensive or aggressive. Territorial instinct shows in it along with the one composite at a time in a place rule behind collisions. Another twist in that, concerning battlefield confusion, is that war's simultaneous destruction and bonding power is based on groupthink sanctions for acts of violence. These are propagandistic in that what would be condemned at the individual level is assigned high value on behalf of the group. This is never more blatant than in the ancient Hebrew commandment not to kill on one's own in the context of repeated commands to kill on behalf of the nation, polarizing non-cult populations as abominations. Thus, an offence punishable in one category is rewarded in another. In the manner of war-making manifestos, sacred texts in general raise groupthink to the highest possible level, in many cases bringing the maker of countless stars and galaxies into the local fray.

War making follows other patterns as well, some of them dictated by weaponry. The trail back to origins can sometimes find a logic of sorts in that - in living things often the survival of the fittest. In a blatant version, nature's having things to be handled leads to hands put to many uses, including combat. Hands line up the sights of rifles, fly the bombers, and launch the missiles.

That the weapons are as made for them as triggers for fingers hearkens back to the freeing of hands for defense and predation in hominid development. That too had a group phase. Nature's creating a need for teamwork in a comparatively fragile biped among stronger predators encouraged tribal grouping and, way down the line historically, brought the nationalism that underlies warfare. The rousing speech of Henry V (*Henry V*, 4.3.18ff) bonds the band of brothers, enlisted against those similarly grouped. In both world wars, patriotic preparation and its widely broadcasted propaganda made virtual powder kegs of nations ready to explode over a minor incident. What interests historians looking for specific causes in such events is much more directly eventful than that, but the us/them polarity in nationalism always makes a contribution.

Again, a pattern (the organization of a population under the us/them polarity) sets off a chaotic panorama. Similar intrusions of one grouping into another go all the way up to merging galaxies. The larger of two galaxies absorbs the smaller one as its gravitational center pulls in stars, moons, and satellites, leaving some areas in nebular chaos. Its conquest gives the total a new overall composite shape.

When Does a Pattern Cease To Be? The Strata Example

Something in motion and constantly changing raises a universal feature of pattern perception: its dependence on the moment and the angle of vision, as moving a small distance changes the landscape an ant sees and moving a great distance undoes the big dipper that people accustomed to long handle dippers see. This, and the samples I've cited, raise the question as to how complicated or unstable a configuration in the visible range, or a perspective on it, can get before we consider it random – a question which applies to both mechanical and behavioral examples. Miscellany prevails in utter chaos, even if its area is enclosed and everything in it is strictly law abiding. One way to proceed with in-between states is Melanie Mitchell's in *Complexity: A Guided Tour* (2009), which puts kinds and degrees of complexity under size, entropy, logical depth, thermodynamic depth, and a few other factors. Fractal diagrams are profusion by self- similar replication. The recurrence prevents them from becoming sheer confusion. On computers, the examples can go down or up a fair amount in magnitudes of 10 so long as zooming magnifies the same-shape replication. Too much multiplying of a pattern turns the overall collection into confusion.

A cooling or solid mass can take almost any shape and still be organized, just as circles, planes, and spheres can be warped and pillars broken without completely obscuring the underlying circular, flat, and columnar shapes. The irregularities of the solar system increase with distance from the center, but all parts, from roughly spherical satellites to outer belt rubble, tell a gravitational story. The 79 moons (or possibly more) of Jupiter follow orbits, another case in which multiplied patterns move toward confusion. Unlike comets, meteors, and asteroids, even the rubble of the Kuiper belt doesn't take a completely renegade course, though something knocked out of it can. A large flying fragment approaching a planet may either be drawn into its gravitational field or be flung inward. It can also be sling-shotted completely out of the solar system. It then looks renegade, but from any given point its vector and velocity are subject to calculations of angle and momentum. A large number of renegades would, again, be chaotic by proliferation. One value of stable placement in an observer is that it automatically sets up direct measurable lines to objects. High speed momentum along that line allows bends and arcs but not knotted twists and jagged lines. This is true of the asteroid from another star spotted going counter to the orbit of Neptune; it had quite a journey in straying into the solar system to begin with, but even that oddity obeyed momentum, vector, and gravitational instructions, and an orbit in reverse remains an orbit. Motion also obeys the inescapable placement principle in which leaving one area (breakdown) is arriving somewhere else (buildup). In a molecular, gas, or dust cloud, various motes and particles join with and separate from others in one of the more incidental of configurations. No matter how disrupted a large body of moving objects is, each and every part of it follows laws - the one thing in a place at a time rule, for instance. Placement remains an invariable. Above the atomic level each integral object is a composite following the rules of the four forces, strong and weak nuclear, electromagnetic, and gravity.

We can, perhaps, test degrees of pattern breakdown better if we narrow to one nearly universal configuration with in-between states. *Strata* look suitable for that because layers are abundant in both mechanical and behavioral areas, mainly as hierarchy in the latter. Other schemes for non-sentient materials would do almost as well such as one-way flow, congruent spin in related bodies, fields, networks, lattice works, container/contained, part/whole; indeed, nearly anything we consider regular if not necessarily universal. One advantage of strata is that they often fall within the optical range. We can find them misshapen, inverted, twisted, and otherwise disrupted so long as they don't become completely disorganized. The obvious example is deep history recorded in earth's rock, which can be disturbed and broken and still remain decipherable. In the planet's phase change from molten to solid, gravity was drawing matter into position around the core according to levels of heat. Once the crust became solid it was pushed up at angles and was broken nearly everywhere. Still, those initial levels could often be sorted out chronologically.

The walls of the Grand Canyon expose some 1.8 billion years of what is probably a typical rock planet phenomenon throughout the universe. The less pronounced layers of gas and fluids also show in viscous materials, which can likewise be sorted out by degrees of heat and weight. In nature, the strata come not only mixed like that but in solids that have been bent, folded, and broken. One layer blends into another and can be perforated by seepage. On charts, to restore history, we keep them apart and neatly arranged. The charts in that respect are idealizations for instructional purposes, but they do after all clarify what is actually there in a rough and tumble form.

Most of this is well established but relatively recent in the annals of earth history, which didn't get seriously underway until observations of exposed cliffs and bluffs and the fossil studies of Konrad Gesner (1516-1565). That testifies as to how important stacked rock has been in the overall conversion from various cultural myths to documented planet history, sequentially speaking, in science history, in going from Gesner to Hutton and on to Humboldt and Darwin.

We've glanced at that course on the way to plate tectonics and need only recall here the role that rock strata played in getting the chronology established and the major events dated. How sea life fossils got to high elevations was up to Hutton to propose, and he did so with surprising detail given the state of 18th century geology. As he saw them, mountains were constantly breaking down and being carried out to sea as silt and sand. From there, new mountains pushed up: "all the strata of the earth . . . have had their origin at the bottom of the sea, by the collection of sand and gravel, of shells, of coralline and crustaceous bodies, and of earths and clays, variously mixed, or separated and accumulated" (13). What tumbles down or blows off the heights complicates, but doesn't erase, the strata. The entire process is heat and gravity powered. The uplift of surfaces into mountains took much more time for earth history than Hutton and his contemporaries were ready to concede - a long time indeed for the sea bottom to rise to the top of the Himalayas.

Hutton's theory was exceptional for the times, in that it required no explanation other than what nature itself contained: "nine tenths, perhaps, or ninety-nine hundredths of this earth, so far as we see, have been formed by natural operations of the globe, in collecting loose materials, and depositing them at the bottom of the sea; consolidating those collections in various degrees, and either elevating [them] . . . above the level on which they were formed, or lowering the level of the sea" (13). As the first methodical geologist in English, he based those observations and his theory on a combination of mineralogy and strata. Two centuries after Gesner, his and other evidence produced still more detailed theories of earth history in Jean-Baptiste Pierre Antoine de Monet, Charles Lyell, Chevalier de Lamarck (1744-1829), Erasmus Darwin (1731-1802), and Humboldt. The continuity of animal and vegetable kingdom lifeforms followed in due course, their history likewise recorded in the strata. In that stacked form, some four billion years of buildup and breakdown proved to be at least partly retrievable.

One moral is clear from this: what some patterns are 'for' is the information they make visible. We seek them out for what they say about the shaping forces and extent of the history. When the archives come in rock, investigation goes to sufficient depths to put alternative earth histories in focus as cultural artifacts. That, too, makes sense in collective us/them terms.

Alternative accounts of the universe were invented, as we've been seeing to solidify populations under priests, kings, and their presumed supernatural contacts. Hence, it isn't just earth history we would not understand without the record in rock, but how cultures and their groupthink work. The distinction between what is actual and what imagined is seldom as decisive as it is when stacked in layers of rock.

Hierarchy

As a common pattern, strata have another advantage in sorting out into hierarchies among flora and fauna. That goes as basic as large dominating small. At the bottom of the heap is where few creatures want to be. At the same time, being too domineering in some pecking orders raises envy, generates antagonism, and incites rebellion. Rivalry and envy show up in other parts of the animal kingdom, but especially in the principal symbol user. Asserted status among people also triggers parody behind the back of high dignity; on stage, a clown struts and swaggers behind a uniformed authority figure. Mockery and overturning follow close behind dignity on display, as it did in the overthrow of high ceremony regimes in the Protestant wars. In militarist regimes, putting uniformed ranks on display in marching formations six to ten across is a common means of dictatorial power display. Not necessarily the parody, but the envy seems to be the case wherever a pecking order becomes prominent. Where bluffing is part of that, the puffing up becomes a way to avoid unnecessary injury. Both a cobra confronting a mongoose and a sniffing bear rising up to assert itself use elevation to put prowess on display. Even botanical kinds have something similar to ascendance and competition (not, of course, the envy) where big trees claim the sunlight and root space that lesser ones would otherwise have.

Nothing in nature assigns species actual rank, but almost instinctively we rank eagles above Insects, and humans above everything. Evolving later usually means superior. When Yeats concerning the frenzy of "Nineteen Hundred and Nineteen" turns zoologist for a moment, he automatically adopts that convention:

> We pieced our thoughts into philosophy, And planned to bring the world under a rule, Who are but weasels fighting in a hole.

Social ranks are less distinct in democracies than in monarchies and oligarchies, but going completely without a hierarchy would show why *anarchy* is *hierarchy*'s 'leaderless' first cousin. Positions and offices in business and government are numerous beyond counting. Organizations couldn't function without them any more than the military could without rank. Nearly everyone who has a job also has a boss, and the boss, in turn, a boss. In Shakespearean drama, when a king goes awry everything suffers on down the line.

Command authority is the customary way to actually put rank to work, as the Latin for *word* is reflected in *dictator* and *dictatorial.* The common way to make rank easily recognizable without its issuing commands is to dress it symbolically and give it a title and a stylized manner, or *posturing* as those lower in the hierarchy are likely to call it. Both civil and ecclesiastical hierarchies have their variants, as does the military, noted for protocol and displays. The more intangible the basis of rank the more impressive the verbal formulas and displays are likely to be. A salute has the dual function of acknowledging rank and showing an empty hand, without a riding crop or weapon in it. Nor is the salute's stiff, upright stance used to spring at anyone. For novices getting initiated into ceremonial orders, indoctrination decodes such gestures and emblems. A large segment of civil and home education concerns manners, a behavioral way to show respect for someone's position in the social order.

Contemporary civil ceremonies, even in democracies, retain vestiges of older hierarchical displays in anthems, pledges of allegiance, and oaths of office. Indeed, next to the issuing of commands, staging something solemn is the most common way to project a ruling order in a readily perceived, participatory way. Repeating solemn phrases indicates obedience to the realm, even the coin of which can have something of high standing stamped on it, as in putting "in God we trust" on something decidedly ungodly. The display must be nonfunctional in pragmatic terms to make its symbolic status stand out. The treasury would work the same without stamped images and phrases. The economy would have the same harvests and GDP and meet with the same setbacks. A good many relics of that kind linger among modern civil and religious organizations, and in social organizations as well. They will probably continue, because collectives instinctively turn to high-dignity display in order to coordinate their ranks. Citizens ordinarily do honor such ceremonies and titles, which are given additional prominence when they are accompanied by monumental buildings adorned with bronze plaques and statues of former leaders. By means of speech and bearing, manners do something similar in casual society, as do marching bands in playing national anthems. The Nazis' extravagant marching style gave visual testimony to the collective strength behind it. The more absurd the style, the more clearly authority has been imposed on it – also, the more likely to be parodied.

One noteworthy departure from social organizations with durable ranks Carol Meyer (2013) calls the "heterarchy alternative" (196), prominent in ancient agricultural Israel outside priest and king-dominated Jerusalem. Modern authority, too, becomes transferrable where it goes by the job in either a temporary or a lifelong division of labor. A technician who comes to fix a computer momentarily has priority - so does the one in India who takes control of a computer's cursor in France - and so on, through hundreds of occupational specialties exercising priorities that adjust authority to function. Teamwork can shift priority in the midst, as in basketball hands on the ball control a set play and determine the shooter, who in turn becomes prime for a moment.

Democracies lean toward heterarchy, monarchies toward despotism, and oligarchies toward multiple ranks arranged in a dynastic hierarchy. Parades with a queen for a day on a floral float suggest that it is almost genetic for people to set someone on a throne surrounded by attendants. Ridicule gives a distinctive human twist to variants as in the jester of the Commedia dell' Arte and jokers in decks of cards, 'take downs' of the pompous. In cards, the joker stands apart from the hierarchy that goes from the deuce to the king and queen. The rules of the game determine its function. On a historical (not a play) level, what were once the actual royal progresses of crowned heads have largely become scenic play acting in modern shadow monarchies. In public, the dukes and duchesses, lords and ladies, kings and queens, are chiefly ceremonial, putting hierarchy on display without giving it a serious ruling function. A queen with a characteristic hand-wave still has the capacity to cheer up a population. So do archbishops with ceremonial moments, colorful robes, and tall hats.

It is harder to find a niche in larger orders and national melanges than it is in communes and tribal hierarchies. The social layers in such orders develop soft edges. Robert Storey (1996) puts "the larger spheres of contemporary life" in unflattering terms, where it is the numbers and not a defined social hierarchy that downgrade individual worth and bring a sense of personal sinking: "Add to this sense of injustice a diffuse fear and guilt, born of the depersonalization of our social exchanges and the weakening of our ties with family and friends, especially with parents and siblings, and the familiarly Kafkaesque angst of our age begins to find its focus. Emotionally and psychologically, ours is the disquiet of the late-Pleistocene hominid thrust into a crowd of alien bipeds-on-the-make and dependent on the kindness of strangers" (60). A sergeant in a platoon under a lieutenant knows his place. A commuter in a Tokyo rush hour, in the vicinity of nearly 40 million people probably knows no one close by, has no definable status in the crowd, and maybe nothing very definite once the commute is over.

Milling about makes for odd momentary combinations in loose orders that are generally too fleeting for methodical study. With patterns that unstable, we settle for blanket terms that cover plazas filled with people and take no notice of connections among adjacent pairs and small groups. That's one place where, when method meets miscellany, miscellany wins. The audiences in town hall meetings being addressed by someone are a seated version of that commingling, focused for the time being on someone elevated on a stage or dais. Crowds can also sometimes be sorted out into clusters where one distinguishable set reacts to another, as in *Coriolanus* patricians control common Romans in bunches not through an Antony-like speech but by marching in noisily, accompanied by attention-demanding drums, trumpets, and oratory "splitting the air." As such groupings suggest, calling a social structure a pyramid is misleading where a dynasty, economic privilege hierarchy, crowd, or an organizational chart have no particular geometric proportion. Physical, biological, and social sciences are wise to leave such mingled confusion to playwrights and novelists. Along with movie makers and television directors, novelists can single out someone in a crowd and string a plot line through the confusion. The crowd is to the story in that respect something like a nebular region to the stars taking shape within it. The degrees of organization and confusion between utter chaos and a recognizable pattern are themselves something of a confusion.

In a corporation hierarchy, it isn't unusual for higher management to receive a hundred times the compensation of middle management, which is less a financial pyramid than an obelisk.

Where one thing rises another sinks comparatively, and where it rises abruptly another sinks precipitously. That geometry is the opposite of mingling crowds and yet not the same as durable dynastic ranks either, though peasants have always been decidedly lower in rank than nobles. The castle/hovel, lord/serf, and mansion/shack disparity has characterized more than master/slave societies, monarchies, and oligarchies. Only utopias manage to avoid it altogether, and they are imaginary. However, as populations grew in mercantile and trade systems, the ranks multiplied to the point of mingling. What had been a dynastic system moved toward heterarchy, a sort of hierarchy dispersed into dozens of suborders in homes, businesses, governments, schools, clubs, teams, and so on.

To Recapitulate: Buildup/Breakdown

One of the remarkable cataloguing and classifying undertakings decades before evolution was recognized was that of Humboldt and his partner Aimé Bonpland in South American in the early 19th century. Another in North America was David Douglas lifting some two hundred previously unrecorded Northwest species out of native and immigrant terminology and assigning them to botany's classification system. As a recent example of imposing an indexed inventory, a century and a half after Douglas when the cataloguing of flora and fauna was nearly complete, a manual called *The*

Weeds of the West, assembled by botanists in 1996, listed over 1500 plants from absinth wormwood to zygophylla-ceae. They were put under the general category *weeds*, meaning useless growth. Both stone-age lore and immigrant terminology had neglected to assign them their torsos and limbs and a capillaried, crenated, clavated, spatulated, tuberous, globrous, lanceolated terminology.

The proliferation of patterns in wild nature doesn't always lend itself to distinguishable categories like that, but the building block level is always in place. The continuum mandates that, until entropy is complete, the laws of atomic matter will keep imposing its rules, hence its patterns. Whatever complications social orders generate, atomic matter and biology set the constraints.

Nature doesn't fall into strata overall, but does always follow buildup of composites and their disintegration, a natural history master pattern. In the biological or lifeform branch, as living composites all members go through roughly similar buildup stages. Every pattern must have something that holds it together, and that something eventually gives way. Even dark matter and dark energy, with no known interior organization, have visible gravitational influences on galaxies. A chart of everything in total makes the current profusion/confusion look to be in apple pie order:

Entropy will bring an end to any further gathering and dispersal with no more patterns in the making and unmaking. That's in lieu, however, of possible geometries and movements that require astronomers to plot not only the vectors and intersections of heavenly bodies, but their long-range movements as a whole; should gravity, for instance, overcome the current dispersal and pull the galaxies back into a big crunch. The possible geometries differ fundamentally from spherical to flat to convex or saddle-shaped. At this stage, nothing can be definitively ruled out - not with 95% of the chart acknowledging dark and unknown matter and energy. Meanwhile, relative sizes and movements, brightness or luminosity, cluster configurations, systems of various dimensions and movements, stages of buildup and breakdown, and other shapes and measurements give students of patterns much to do. As the forgoing chart indicates, that's with visible matter adding up to only a small fraction of what gravity indicates is out there. The rest escapes visible embodiment and any portrayal of patterns other than what massive black holes impose on the luminous bodies around them. Like Saturn's rings, the mix is chaotic but the orbits are semi-geometric.

CHAPTER EIGHT

CONNECTING FORCES AND THEIR LIMITS

The Argument

The details of how composites come together and decay fall separately to the disciplines. In putting these in deep history, natural philosophy stands apart from the inner workings, but applies the determinism of the continuum to them at the building block level. Up to creatures that make choices, the determinism of the continuum operates strictly through atomic binding and breaking forces.

< >

Out of the tomb, we bring Badroulbadour, Within our bellies, we her chariot.
Here is an eye. And here are, one by one, The lashes of that eye and its white lid.
Here is the cheek on which that lid declined, And, finger after finger, here, the hand,
The genius of that cheek. Here are the lips, The bundle of the body and the feet.
. . . . . .
Out of the tomb we bring Badroulbadour.
(Wallace Stevens, "The Worms At Heaven's Gate")

< >

What Does the Binding and Breaking

Again, I'll necessarily rely on sampling to get at the key kinds of bonding and breaking. I've settled on Stevens' Badroulbadour as an especially critical case at the human, biological level - critical because rather than having her disintegrate all the way down to the soil's particles and being converted eventually into plant growth, Stevens puts her in the context of the resurrection story. That she hasn't fully disintegrated reminds us of who she is and what sort of life she led. That lets naturalism and religion clash as they will, as they do in one way or another in much of the world's population. The reader has to choose between Badroubadour as a representative of the continuum and an alternative, typical of human oriented schemes, that defies atomic matter invariables.

The succession of buildup/breakdown composites under the naturalist umbrella has a limited number of connecting forces responsible for the patterns sampled last chapter. The notions of the chief symbol user, equipped with language, math, and visual means of assembling and disassembling, venture frequently into concepts only loosely connected to reality. Within the familiar sensory range, ways of capturing what has broken loose and making something different out of it are commonly known, as a rabbit is disassembled to sustain a coyote and abattoirs carve up pork to sustain humans. We can set that aside with merely a reminder of how universal such recycling is. Everything made of atoms undergoes it. A good deal of the universe remains in that uncoupled condition, in free-ranging neutrinos and hydrogen and in the separating galaxies.

Getting down to the most compact and strongest of the cohesive and breaking forces was a formidable task, initiated nearly a century before Heraclitus by Thales of Miletus (about 624-546) and, at the philosophical level, taken up by the Greek atomists. Among the pre-Socratics, those who turned to atomism had the atoms joining together and breaking apart accidentally. Of course, they hadn't actually watched that happen, but it would explain the large scale disorder of nature. The recurrent patterns required knowledge of natural laws that weren't known until comparatively recently. Investigation into micro levels stalled after Lucretius, understandably, since without magnification it had nowhere to go. Nor could investigation into most of the celestial mechanics proceed without telescopes to reveal the macro level. Even as late as the 17th century, some two millennia after the atomists, the available telescopes and microscopes weren't powerful enough to catch anything as small as a molecule or to identify anything as distant as a galaxy.

We can gauge the extent of what observation, theory, and philosophy have produced since the introduction of higher powers of magnification by setting a representative earlier world view alongside something modern. Spot checks of that kind wouldn't tell us a great deal if there weren't representative figures, but we fortunately have several to choose from in the alternative histories to which Stevens' supreme fictions probably refer. Contemporary with Copernicus, Brahe, Kepler, Galileo, Bacon, and Newton, the dominant view of the cosmos in the Middle East and Europe was a variously combined Greco-Roman and Judeo/Christian one. Bacon was ahead of his time in England in setting that combination aside as a field of research and arguing for induction. Like Copernicus and Galileo as proponents of the heliocentric model, he too believed that what observation turned up would be found to agree with scripture, since the authorship of both was the same. The common assumption was that the universe had gotten underway a few thousand years earlier with a series of commands and immediately turned the planetary part of it over to human management. One day not far off, it would end in a fiery apocalypse that would destroy flora and fauna and send non cult members off to eternal torment. Threatening that catastrophic end had become a powerful cult recruitment beginning over 1600 years earlier.

Under conventional Christian doctrine, to be forever burning would be far worse than the previous sacred order's genocidal campaign in eliminating non cult abominations, beginning with Canaan and extending to much of the eastern Mediterranean and Middle East. Human error, amplified by divine wrath, explained rugged terrain, climate extremes, predation, disease, and everything else not right with the world. Both blessings and curses were intentional.

Early 20th century modernism combined too many factors to make anyone or any movement quite as representative as almost any earlier version of the story Milton put in epic form, but what had happened to older cosmologies and world views by the 1920s is clear in Stevens' reclassification of them as fictions. He didn't specify any particular one, but since *Paradise Lost* tells the most orthodox of the stories in the West from beginning to end it serves admirably.

Generally speaking, to that way of thinking nature and errant people were assigned the decay and breakdown phase of the cycles, leaving the restoration to miraculous means. For people individually, the key moment comes at the judgment, specifically at heaven's gate where lives are summed up and their future determined. In "The Worms At Heaven's Gate", cited in the epigraph to this chapter, Stevens humorously assigns the worms the first phase, collecting and transporting Badroulbadour's once well assembled parts there. Their arrival underscores the problem of the next phase. On their own, the talking worms obviously can't put their cargo back together, fine specimen that she recently was, to yield a result worthy of a poet's blazon. (The poem, in one guise, is just such a poem of praise, recited too late to catch Badroulbadour at her best.) Stevens stops with the worm's narration and doesn't proceed to what comes next - though "Of Heaven Considered As a Tomb," "A High-toned Old Christian Woman," and "Le Monocle de Mon Oncle" go a step or two further. Elsewhere in the text, the angels that ride down from heaven on mules bring a similar clash of opposites and drop in tone, falling off in grandeur comparable to Captain

Stormfield's, arriving at the same place still intact.

Recycling bodily parts could be left to nature, which knows how to do that much, but *reassembling* specifically Badroulbadour requires an agency that, if it had that kind of skill and power, presumably needn't have disassembled her in the first place. In naturalistic terms, to undo the damage from death and decay requires that every cell and microorganism among trillions has to be revived or matched, assuming that the result duplicates what once existed, if Badroulbadour herself is restored and not an idealized or ghost version. Cloning might manage that degree of rebuilding, but it starts with a sample and has nature take it from there. Frankensteinian fiction puts a comparable task in the hands of a semi-mad genius, but, like Stevens' poem, that fantasy too leaves the stitches showing.

Remembering the former life is necessary for a restored individual psyche, and it requires the reconstruction of the particular mind. Any revival of a point of view goes beyond individual reconstruction and includes what families, congregations, books, and cultures have imprinted on the mind. Then, a setting has to be provided with her personal contacts and theirs in turn, and so on through the population Captain Stormfield encounters. That's many billions - perhaps hundreds of billions if the human race lasts long enough. Imprisonment in perdition for the majority doesn't solve the problem, because the condemned, too, have to be reconstituted to undergo unending torment and remember what landed them there. What is also usually left out of a conventional resurrection are age differences and hierarchies that complicate living relations. In some beliefs, for instance, paradisal subdivisions are headed by patriarchs without accounting for how many senior patriarchs that would generate, where those who were once junior eventually became senior. Each generation is connected to former ones, resulting again in billions of senior patriarchs with no junior ones.

How does the tradition handle such logistic problems? Generally speaking, by ignoring them as Milton does and as the mathematics advises. George Herbert is typical of Christian variants in lining up stages and levels of removal from individual death through universal destruction and on to reconstruction. In the apocalyptic torching of the world, mankind will be "out of order hurled,/ Parcel'd out to all the world" ("Doom's-Day"). In "Decay," souls but not minds and bodies prepare for that by scouring themselves clean beforehand. Scouring clean means basically conforming to cult standards and going through its rituals and sacraments. These get their efficacy not by anything inherent, but by divine means assigned to one given denomination. The conflagration when it comes is understated in typical Herbert manner, in a plain style that makes it sound simple and painless:

I see the world grows old, when, as the heat Of Thy great love,–once spread,–as in an urn Doth closet up itself, and still retreat,
Cold Sinne still forcing it, 'till it return, And calling Justice, all things burn.

The final line is a marvel of compression, combining justice with righteousness – justice because cold sin has grown intolerable but, at the same time, odd, because it assumes that fish, plants, and animal kinds ("all things") and the good and the evil are alike worthy of destruction. The heat "Of Thy great love" is an oxymoron in combining *love* with destruction by that unloving inferno. The monstrosity that animating and moralizing nature creates has once again taken over. That in most end-thing visions the creatures aren't awarded an afterlife presents another problem in logic and justice. The failures of the upper hierarchy, angels and humans, pass down to the animals, stricken with disease and mortality, but the sorting out and triumph of the virtuous doesn't draw them up.

They suffer the penalty without the redemption. The fate of infants and children also slides as quietly past as it does in the mass drowning of the Noah story, the latter a clearing out of lifeforms, and bringing a new start for a small boatload of survivors followed by a post flood human history as bad as the one that brought the flood. (Under conventional omniscience that was foreseeable, thus the futility of the mass killing.)

The conclusion to Milton's youthful "At a Solemn Music" goes directly to the end state without the conflagration:

O may we soon again renew that Song,
And keep in tune with Heav'n, till God ere long To his celestial consort us unite,
To live with him, and sing in endless morn of light.

In the context of *Paradise Lost* (still years away), Milton is typical in joining the New Testament concept of an end to the Hebraic beginning and some Greek cosmology. He also adds a way to keep the numbers down, since the final consort will exclude those whom Satan and his cohorts have seduced. In the post-Restoration writing of *Paradise Lost* and *Paradise Regained*, that might have looked like the majority - at least of Englishmen. The glory of the celestial consort momentarily escapes the boredom Stormfield finds. It could return if the only thing to do forever is sing. In the abbreviated history of the world assumed in the 17th century (less than one presumed year for about two and a half million actual ones), that would make the numbers almost manageable if not reduced enough to allow the customary personal reception and judgment.

Easily overlooked in Milton's abbreviated version is the removal of actual sun/earth relations. In the rewriting of natural history, light is fixed in an everlasting morning, which finesses darkness as "ere long" finesses the very long wait, already over some 17 centuries past the promised date.

Another alternative occurs to Marvell in the neoplatonic framework of "On a Drop of dew," the dissolving of the soul itself into the Almighty Sun, clearly not the usual one. That would shed not only the encumbrance of a body, but also individualism, subjectivity, memory, and matters intellectual. Because that version of the soul dissolves as dew drops do, it doesn't matter that nothing of the individual is left. Such notions run into difficulties for the obvious reason that they don't fit what natural and human histories provide.

Prospero's version of the end in *The Tempest* is slightly more naturalistic, but still unrealistically placid concerning eschatology:

> Our revels now are ended. These our actors, As I foretold you, were all spirits and
> Are melted into air, into thin air;
> And, like the baseless fabric of this vision,
> The cloud-capped towers, the gorgeous palaces,
> The solemn temples, the great globe itself, Yea, all which it inherit, shall dissolve, And, like this insubstantial pageant faded, Leave not a rack behind. (4.1.148-156)

Only if existence were a dream could it fade away that easily. Prospero is ignoring the difference between the spirit actors he has summoned for the pageantry and the real people standing before him – or, to put that another way, Shakespeare is ignoring the human actors enacting spirits in front of a bodily audience. Neither the actors nor the audience could dissolve casually any more than Herbert's creatures could burn painlessly. The goal is clearly to be rid of nature's flaws once and for all, even if the means is turning Everything back into Nothing.

One further example gets us closer to the modern end of the Renaissance-to-modernism span.

Consider Robert Frost in "Fire and Ice" reducing the apocalyptic end to what people have somewhat casually remarked:

> Some say the world will end in fire, Some say in ice.
> From what I've tasted of desire I hold with those who favor fire. But if it had to perish twice,
> I think I know enough of hate To say that for destruction ice Is also great
> And would suffice.

Frost's interest isn't in the end itself, and certainly not its mechanics, but in the psychology of burning desire and freezing hate among the living. What may exist somewhere else isn't among his speculations, which, in any case, in "some say" seems little more than a rumor.

Other accounts of large-scale disintegration could be cited, from the mystics' dark night to Donne's "A Nocturnal," his two anniversaries with their "just destruction of this world," and on to Yeats at the 20th century's end. These are enough to show why, in a modernist framework, Stevens considered traditional alternatives to be fictions, supreme ones if they were ambitious. The modern alternatives in atomic terms are several, not just the eventual quiescence of burned out stars and total entropy. In "Could Time End" (2012), George Musser lists several derived from physics that avoid Stormfield's population and the worm's reassembling problem, a big crunch, a brake, a lurch, a rip, a freeze, and a whimper, the latter owed to T. S. Eliot's "Hollow Men." Those alternatives aren't the only ones in the unforeseeable future. String theory adds others that Brian Greene (2011) classifies under kinds of multiverse, "Quilted, Inflationary, Brane, Cyclic, and Landscape" (190). The rules that apply to these likewise belong to physics, chemistry, and time's relentless forward movement - another unavoidable constant.

It may or may not be relevant to recall that Stevens' concept of supreme fictions ran parallel, in the earlier 20th century, to the framing of the natural continuum. The 1923 date of *Harmonium,* a collection of many years' work, coincided with Hubble's sighting of galaxies, begun in 1919 and announced in 1924. That discovery changed the course of subsequent astronomy and was soon matched by estimates of the distances the galaxies had gone in their outward flight, with their velocity gauged by the redshift's receding light waves that show the farthest galaxies to be moving away the fastest. The astronomic realm paralleled revolutionary discoveries in the subatomic realm of protons and neutrons, which also made significant progress in the 1920s. In that context, Stevens' first volume did something similar in a branch of poetics shared with other early modern figures, among them Yeats, Frost, T. S. Eliot, Ezra Pound, W. H. Auden, Elizabeth Bishop, and Marianne Moore. These, and others, also sometimes ventured into world views and philosophy, albeit more incidentally than Stevens. Every field, it seems, was producing new models, from quanta to cars and refrigerators. When the standard model did finally get formulated, it supplied a better-defined bottom level, including the breakdown of electrons, protons, and neutrons into their particles and binding forces. The dismembering of Badroulbadour is a stark reminder, at the human level, of that breakdown phase, but in buildup/breakdown principle it isn't essentially different from any other atomic matter composite. If we put subjectivity entirely aside, our personal disintegration becomes just another instance of innumerable universal breakdown cycles.

Powering the Cycles

Since no pattern is possible without a bonding principle, this chapter is a close companion of the last, except about what produces and disassembles patterns rather than about the configurations themselves. We look for binding and breaking forces at all levels, in all fields, from clues in a murder mystery to what propelled the big bang and the brief moment of heated inflation. In buildup, the combining of the smaller in the larger can be accompanied by the butterfly effect, in which a slight difference between two things at the outset escalates unexpectedly. A butterfly flaps its wings in Brazil and a tornado roars through Kansas. What the bonding joins can last an extremely brief or an extremely long time; ten billion years isn't unreasonable for a star the size of the sun. The most powerful level of attracted and bonded matter comes in black holes, leakage from which is presumed to be what lasts the longest. Generally speaking, the stronger the bonding force, the stronger the breaking one has to be. It takes great heat or a high velocity collision to break atomic nuclei bound by the strong force. Nothing extraordinary is needed to overcome gravity for a moment at earth-mass strength. Chickadees can accomplish that with spring in their legs and wing beats. In a neutron star, once gravity as the cohesive force has overpowered energy, further atomic level disintegration is forbidden. As distinct from the bound neutrons in them, the stars themselves can collide and send off gamma rays.

What the quanta are and how they come together and separate were among the last phases of atomic science to be understood. What binds an electron to a nucleus doesn't always hold, nor does the nucleus stay intact forever. If the cohesive forces *were* consistently overpowering, one atom wouldn't surrender its components to another, the sun wouldn't shine, and we wouldn't be here. Wreckage at a supernova level produces a high level of disintegration, but the resulting fragments can reform as they did in the solar system. As for lifeforms, their most severe die-off and recycling came at the Permian/Triassic border, one hypothesis which holds that the microbe *methanosarcina*, an anaerobic methanogen (methane producer), went viral in converting carbon into methane. What *Wikipedia* says in chemical terms is that the methanosarcina, which under a microscope look like harmless fluffy balls, were the product of a metabolic pathway enabled by transfers of genes and their "exponential reproduction." That looks like an abrupt change, but again the components were only following necessity. A tornado that seems an excrescence is no more rule breaking in that respect than a sea breeze. The trashed structures will work into something else in due course. Things come together. Things fall apart. Acknowledging that gives us something like ‘law-abiding confusion’, which, if we start with natural processes and not with something intended, isn’t the paradox it seems.

The list of tangible and intangible means of cohesion among objects is extensive, as are terms for the results, among them *ensemble, aggregate, bunch, crowd, conglomeration, patchwork, network, hodgepodge, matrix,* and *assemblage,* a set of collecting and bunching terms we’ve watched McLaurin acquire one by one. Because bonds in the animal kingdom are partly affective and mental, they multiply the number of agents and patterns. Beyond those come the prolific notional bonds of the symbol user. The ideas and emotions that people add to the connecting forces aren’t intangible, merely quixotic, prolific, invisible, and elusive. Every gadget in a household, every motor that runs, began as an idea. Digital geometry multiplies the configurations into thousands of nodes and connecting lines that, when colored, are as spectacular on a small scale as astral bodies on a large scale. Objects produced by hands-on manipulation add *contraption, convention, concoction,* and *contrivance,* to proceed to just the letter C. *Conviction* and *concept* add two more notional connectors.

I glanced earlier at the prefix *dis* for some of the breakage, *disintegration, disagreement, disobey, dishevelment, dissonance,* and *dismember. Disable, decompose,* and *decay* were preceded in Badroulbadour’s case by *death* and possibly by *disease.* The *D* list begins with the severance of one thing from another, then proceeds to more general disassembling all the way up to the on-going dispersal of the galaxies. In the breakdown phase of integral objects within our normal range, chemistry performs the final stage using rust as a visible means. The most powerful breakage engineered by people so far has been that of atomic bombs - unless we credit warming seas for an increase in hurricanes and other storms. To put that in focus in terms of scales, Mt St Helens blew its top with a force estimated at 33,000 Hiroshima atomic bombs, and it was small compared to solar eruptions on display in the SDO gallery. Charts of nuclides, isotopes, the elements, pamphlet and wallet references, and other condensed data sheets provide convenient access to what the mechanics produce. The forces behind them aren’t themselves visible, but modern equipment can see and measure what they are doing. Electron microscopes that see atoms, and even one that feels them, are now capable of that level of observation.

Until the invention of the atom bomb and nuclear reactors, it was mostly physicists who thought that finding the bottom layer of atom making and unmaking would lead to pragmatic applications. That changed in public perception in the mid 1940s, when the atomic sciences spawned not only weapons but additional new industries and improved medical imaging.

Chemistry figures in much of the bonding as well as the disintegration, to a large extent by manipulating valence electrons (those on the outer electron fringe of atoms). These are the ones most often involved in binding one atom to another and thus in the forming of molecules. A chemical element can thus set forth on its career, timed in rate of decay but not necessarily in life span, since it can meet with contingencies that break it down ahead of its standard programming,

The flurry of subatomic activity in the earlier 20th century created several new branches of science: the measuring of photoelectric effects, combinations of quantum physics and relativity, quantum electrodynamics (QED), and nanotechnology, to name a few. With the latter came not only the capacity to assemble and take apart extraordinarily small things, but X-ray penetration of materials. Radiation throughout the universe shoots rays into and through bonded atomic matter, sometimes working changes as laser beams do in laser surgery. The impact of the breaking force needn't be dramatic. At the level just above elementary building blocks, atomic bonding underlies molecular chemistry with its non-covalent, reversible and irreversible covalent, ionic, and other cohesive principles essential to life. In ionic bonding, electrons that transfer from one atom to another unite two or more atoms. (Linus Pauling's 1939 textbook *On the Nature of the Chemical Bond* is a historical landmark concerning that level of bonding.) Proteins, nucleic acids, carbohydrates, lipids, and other molecules do the lower-level binding, rising to visibility in tissue. Chaos theory, too, applies to certain bodily functions that have become jangled, to atrial fibrillation and heart arrhythmia, for instance. That's at the tissue level, but, like all bonding and breaking, it starts smaller.

Some chemical combinations are stronger than others, but each blend of elements has a melting or breaking point. In disintegration, the elements and molecular bonds lose their orientation vis-a-vis one another. This is a reasonably apt definition of breakdown, since what both sentient and non-sentient disintegration amount to is components going their separate ways. On the macro scale, disintegrated stars are among the more remarkable examples. For some time, it has been possible to watch spectacular explosions in the unmaking and proto stars in the making. The molecular clouds and nebulas that result assume many shapes, and are the largest optical effects of disintegration. The Cat's Eye, Orion, the Swan, and the Horsehead have turbulent pillars extending to light years, even just one of which would be over 30,000 trips to the sun and back.

Every patterned shape except clouds of dust and gas, however irregular, has a cohesive force at work, electromagnetic and gravitational across distance. Even dust clouds have some gravitational bonding, albeit weak and ineffective outside the denser parts, nodules, and whirling disks in which proto stars and planetesimals take shape. New formations begin that way before proceeding to 'pebble accretion' and continuing to bodies large enough to have their own gravitational attraction. Not every stage of that accretion is well accounted for - as, for instance, in the dense cluster labeled Sagittarius A-star near the center of the Milky Way. How did accretion become so dense in that region? The gravity of a massive black hole, we're told, is one possible answer. Electromagnetic celestial storms do some of the disintegrating as well as some of the initial bunching.

All told, whatever bonding and breaking principles make an inclusive list, we can be sure of the following:

- For the same reason that a distance can't be crossed if it is infinitely divisible, matter must have minimal components - even if we don't know what they are. These cannot be broken.
- Combining smaller into larger units needs not only a binding force, but readiness in the segments to be joined. The latter is a material property, like the capacity to repel and attract in magnetic poles. Protons and neutrons must be subject to the strong nuclear force within a limited spatial and temperature range. Masses must be subject to gravity, the strength of which has been calculated at about 10^{38} weaker than the nuclear force.
- Under the salience principle, some parts of a composite are more important to its integrity than others, but strength and importance aren't necessarily the same. Proximity contributes to salience and applies to both connecting and breaking strength.
- Objects collide if they try to occupy the same space at the same time. The forces (bosons) don't occupy space and can pile up. Fermions (matter) do, and therefore never coincide.
- The breaking of composites can be by inner fracturing, intrusion, or outer flaking. Topographic erosion is outer flaking, tectonic plate movement inner fracturing. In social orders, migration and exile are flaking, revolutions inner fractures, and invasions intrusions. In living things, flaking can discard something that gets replaced in kind, as a snake replaces its skin and trees replace patches of bark.
- Sections of planet earth break and recombine under all three means: fracturing, intrusions, and flaking. Biological, botanical, and

political/social bodies do so as well.

Salience in Mechanical and Notional Bonding/Breaking

What the repercussions of something might be are, at times, as unpredictable as what brings down an avalanche. That the heat accumulated in methane has over 30 times the warming influence of CO_2 isn't what anyone is likely to think about when starting a gasoline or diesel engine and driving off; several million have a cumulative effect intended by none of their drivers. Consequences that land where they aren't expected can also rise or fall in scalar level. Discarded subatomic particles and rays in one place become green growth in another. One effect of that is the difficulty any given discipline has in tracking a succession over an extended length of time. Since neurology isn't evolutionary biology, which in turn isn't zoology, the sciences have no diagnostic procedure for determining what happens stage by stage over the long haul. Despite a good deal of crossover, the disciplines independently see their sections of the continuum like the panels of a stained glass window.

Even just human-originated buildup/breakdown surpasses any means of keeping tabs.

Electronic filing and libraries full of books bear testimony to that. In electronic devices, words, dates, alphabetical order, and size can be assigned a summoning command.

Compare that level of virtual or notional gathering to someone getting home from a stadium, where team spirit and mutual interest has done the main crowd bonding. At work the next day, the same individual, no longer a member of a crowd, rejoins another order or several in succession, all the while maintaining a dues paying membership in Kiwanis or the KKK. Mental play and ideas, meanwhile, hook up and break off intermittently. AI can produce additional lists and trace countless connections, some of what they cite held together by actual nuts, bolts, nails, and glue.

No one will ever determine what was going on in Keats's mind when he made the strange contrary-to-fact declaration, "'Beauty is truth, truth beauty',–that is all/ Ye know on earth, and all ye need to know." The *is* establishes a notional bond in flat denial of reality, not a normal procedure even in a form that binds its lines together by uniform length, end rhymes, image patterns, and themes. In reality, truth frequently separates from beauty. In Keats they are forever bound.

Piled up factors can make even limited areas confusing. Salience helps sort that out. What is essential and what is expendable to a given composite goes by degrees and isn't necessarily traceable. A limb subtracted from a tree or from a quadruped leaves much the same thing.

Subtracting invasive bacteria and viruses from an animal leaves it all the more intact. Removing the heart of a quadruped, a queen bee from a hive, or a strip of bark from around a tree, however, is another matter. Changing the head of an administrative structure affects all subordinate departments, an operational parallel to tiers of abstraction and cascading files. Hierarchy and salience combine with structural integrity in living things in ways not applicable to non-sentient things. Loose combinations, such as ecologies, vary in what counts most. Removing wolves from an area, for instance, increases deer and decreases what they feed on, which therefore affects birds and insects down the line.

Transitions from one composite to another can get technical, as when a crippling disease like the fungi-collecting and spreading polyphagous shot hole borer tunnels into a tree, carrying in the fungus. That is bad luck for the tree, as are insects and borer infections pursuing their own line of existence. Most manmade things come with a built-in salience principle - not only the affective bonds of groups but the planned and structural ones of organizations. These can go unstated or be as formal as blueprints. The supporting beams of a building obviously count more than its decorations. The parts in an automotive supply outlet are placed in more or less equal bins but, when installed, make wheels and axils count more than hood ornaments. Manufactured, handcrafted, carpentered, machined, and cooked up concoctions illustrate countless versions of this principle. Again, notions precede real hookups. These are familiar matters, but we don't often link them to the universal principles of atomic matter buildup-breakdown cycles - despite the fact that indirectly, through long avenues of incremental change, everything derives from them.

Take something even more routine but broken free of actual things and consigned to the notional sphere: the term paper assignments of courses. In discursive writing, it is a theme or thesis that collects the details, in other words a generalization, in narrative a story line. Topic sentences provide the notional glue of paragraphs. Prepositional phrases and adjectives can come and go in limited numbers without much effect, but a storyline, a thesis, or a topic is essential. In a monograph, a specified major one has the help of subcategories in organizing the details.

Though these are mental productions rather than material bonds, they often do refer to exterior connections, just as an equation covers many material instances in mathematics. What is bunched under $E=mc^2$ is virtually everything in existence. A connecting concept or thesis can also be entirely imaginary. In the Hebrew anthology, writers from different eras lined up under an imaginary Abraham/Yahweh covenant. (I'm assuming that the maker of a hundred billion galaxies going on 14 billion years in operation didn't actually decide to champion the citizens of a small Middle East nation, meet personally with its leader, and form a covenant.) That isn't the only connecting thread in the series of texts, but it is second only to the territorial possession itself. In that respect, the nationalized deity of Josiah's 7th century 'Deuteronomic reform' functions much like Anu, Enlil, Amur, Marduk, Tammuz, and other predecessors and successors over several thousand years that use similar imaginary super powers to coordinate populations. That has proved to be an effective people-uniting tactic, but the populations themselves are actually grouped much more by proximity, heads of state, cabinets, laws, ordinances, and dynasty continuity than by phantom beliefs or solemn choral recitations. These confirm and bolster the bonds of proximity, the most important factor in nationalism.

What Aristotle in the *Rhetoric* calls *topoi* (topics) provide another common notional binding force. The topics can be mostly mental and verbal or related to something known from experience. Essays and story collections lacking such a convening topic are more like a collage than a single unit. Unlike a testament or a monograph, an essay collection normally has no particular succession and little, if any, cross referencing. Bacon's essays on Truth, Friendship, Death, Adversity, and Ambition exercise internal thematic control, but the collection itself has no development. Even a single essay can be an assortment of loosely related aphorisms. In libraries, no priority comes with a given call number except in a particular search. The single unit, the library, has a unified purpose, the structural integrity of its housing, and a coordinated staff, but its volumes are grouped by arbitrarily numbered categories under general area command, fiction here, history over there.

Comparable to library arrangements are what computer browsers bring forth on command, the key terms being not necessarily theses, titles, or categories, but sometimes only a word or phrase. Unlike a card catalogue, a browser often has priorities of its own that place paid-for entries in the way of what was summoned or alongside it - equivalent, in library terms, to what would be advertisements placed on placards at the library entrance. Electronic data-in-waiting are collectively a miscellany. In contrast, a monograph isn't allowed to stray from its topic any more than a novel is allowed to break into short stories. Story collections are sometimes loosely hooked together by recurrent themes, characters, and chronology, but the units remain relatively independent compared to the chapters of novels. A succession of writings by different authors in which later ones cite earlier ones is looser in arrangement than histories, but more tightly bound than essay collections. In epic, comedy, lyric, tragedy, and pastoral traditions, later writers are aware of at least some predecessors, and make either direct reference to them or allusions that some readers catch and others miss.

We can see why, in defining topics, Aristotle was important to the history of discourse and Thucydides to the rise of historiography. Both found topics capable of collecting and arranging materials without relying on local chauvinism and hyperbole, always distortions of what rational objectivity would sanction. Thucydides reports the different pieties of his subjects without endorsing them himself. Herodotus, his near contemporary, keeps gods and goddesses more prominent on the grounds that policy makers did so themselves, and perhaps because supernatural figures set the boundaries of a historical reconstruction. If Jove, Mars, or Venus is the final resort, no long course of events need be traced. In taking over the seas, Neptune discards most of the physical properties of water and air not to mention the history of the planet. We are so accustomed to these aspects of notional binding that we seldom stop to think how recent and innovative they are in the context of evolution. A graph registering invented as distinct from instinctive means of connection wouldn't take its sharp upward curve until about 50-40,000 years ago, when proto language and basic numbers developed into more manageable language and math.

Teaming Millions

Groupthink has not only geographic boundaries, but also term limits. The League of Nations ends. The Soviet Union separates. The British Empire breaks up. Comparable examples aren't hard to find in other parts of the animal kingdom as well. The breakup of something formerly bonded can be painful, which is the main reason we like to think of coming together rather than parting, of marriage and birth rather than divorce and death. A meerkat that has gotten separated from its mob whimpers piteously. Together with trains of refugees, the declining stage of a human confederation is a regrettable side of that. The last member of the Yana tribe, Ishi (ca.1860-1916), was pitiable when he came to public attention in northern California near the end of his life. The last of the Mohicans made a stirring James Fenimore Cooper historical novel. Exiles and outcasts are a subset of that social breakdown, as in distinguishable groups such as felons sentenced to prison and less ardent Communists in the Soviet Union sent to Siberia.

The more rigorous and oppressive a government, the more likely it will generate rebellion and trigger its own breakdown. That has held true often enough to be a rule of thumb. Autocratic rulers and dynasties face internal factions that tempt them to gather a loyal inner core and apply the resources of a police state to factions. They also sometimes purposefully stir up external hostility to overshadow internal dissension. Sargon's Akkadian dynasty, the first civilized example of that on record, went on a conquest campaign in the third millennium (BCE), only thereafter to get caught up in recurrent wars waged to hold the empire together. This carried on through Babylon, Assyria, Persia, Greece, and Rome all the way to Attila the Hun, Napoleon, Hitler, and Stalin. Banana republics continue to turn out new examples. Such was also the case intermittently with 17th to 20th century European and American imperialism. The American revolutionary war against the British and the colonial wars of Africa and India were prolonged and bloody as was the American 'conquest of the West' at times. Where a cohesive power imposes ideology or doctrine, subunits break off with some regularity, perhaps because inventing an opposing equivalent isn't difficult. Even sensibly assembled and governed organizations can become as fissured as cracked porcelain.

As fascists and communists have illustrated over the past century, ideology is another source of simultaneously binding and divisive hardline rule. Doctrine is equivalent to ideology in that respect, and has an even longer history. Since one goal of a new devotional sect is to convince members of another to switch, it separates itself dramatically from rivals, thus Christians drawing away from Jews, and Islam from Hinduism. Since nothing demonstrable or rational argues for any given such order, its members get their cohesion from faith bolstered by groupthink. When ethnic considerations and proximity are added to the complex, we can get a severely troubled Palestine/Israel or India/Pakistan. Nonetheless, new sects often start as near variants of older ones – for instance, Anglicans under Henry VIII, aiming primarily for the king's divorce and maintaining national independence, incorporated many of the Vatican's sacraments and beliefs. Despite their similarities, the two high church variants were polarized in both England and the Vatican.

Naming the two holy land anthologies Old and New Testaments was, likewise, a way for the new order to establish itself in the context of a tradition a millennium or so older. Adding other-world rewards and extreme punishments marked the main difference. Transferring the promised kingdom off the planet meant no equivalent to the ancient Hebrew (fictional) slaughter of Canaanites.

The proliferation of self-identified groups can also work the other way around. When a good many of them fall within national boundaries, open hostility among them becomes unwise – seen, for instance, in the many beliefs that took root in North American settlements. Each denomination had its own rituals and ministry. That most of the sects originated in Europe and agreed on major tenets helped them coexist. Tolerance was less the case in Europe in the Protestant wars of the preceding centuries, and wasn't exclusively the case in the Americas. Persecuted groups like the Puritans and Mormons sailed away or moved out in order to remain intact. Donald Kagan (1969) cites a classical example in the Spartans and Athenians: "It would be a mistake to underestimate the political importance of religious associations. We shall see that even Athens was concerned to establish a religious basis for her imperial hegemony" (50-51), the goal of which was to reduce the independent status of other city states and make them satellites. Even Athens is appropriate, since Athenians had philosophy, history writing, and a largely secular constitution but continued to make ceremonial tributes to an imaginary goddess. They were following a couple thousand years of incorporating one or more celestial super powers in a given hegemony.

What frequently helps bond diverse populations is collective hyperbole, not always with a piety or ideological element, but based on ceremonial occasions with anthems and pledges of allegiance. Together with the collective psychology of groupthink, a territorial instinct lies at the root of that. Proximity is more instinctive and real than intangible unifying devices. That may seem overly reductive, but the animal kingdom has similar examples. Only by keeping to their territories can adjacent mobs of meerkats hang together and avoid fighting. Tightly cohesive under their respective alpha pairs or queens, they can tolerate rival mobs only if each keeps to its place. The psychology of us/them in other species that form into packs can be attributed to bloodlines along with the territorial factor. As for people, when it comes to what David Barreby (2005) calls our tribal mind, "There is apparently no people known to history in anthropology that lacks a distinction between 'us' and 'others'" (125). Positioning one across from another sets the opposition and locks each into a tighter unit. They can then congregate under flags, banners, anthems, and group hyperbole. The effective bonding agents are territorial, the notional ones are regional and cultural; they are concoctions made to seem as authorized as the sun rising and setting.

Sticking Close

I've been using *proximity* generically to cover variable ranges of distance, from individual core subjectivity and households to nations. In all its spans and degrees of familiarity, placement and territorial occupation is what matters in combining and separating. It is time to be more specific about that, since up and down the scales, from a nucleus with electrons to stars with planets, physical distance determines what applies to bonding and breaking. The range of the four atomic matter forces is well known from the strong nuclear force at work on atoms building up to gravity's universal field, gauged by Newton's inverse strength principle: every body attracts every other with a strength proportional to the product of their masses and inversely proportional to the square of the distance between their attracting cores. The cores themselves shift depending on the strength of the opposite attraction. For carnivores in whom attraction meets with predation, a defended striking distance can drop to that between the mongoose and the cobra. For gangs like the meerkats, it expands to hundreds of yards. For nations, it is thousands of square miles and the range of missiles. An imperialistic power that pushes out its borders crosses into zones that are inherently hostile to it. When Napoleon confessed that his power depended on glory, and glory on victories – that conquest made him "what he is and only conquest enabled him to hold his position" (Paul Kennedy, 1986) – he was, of course, sweeping alien populations under French rule who didn't want to be there. This has been a recurrent story of dynasties and empires for several thousand years, in the process breaking down proximity bonds by predatory force. Nearly everyone belongs to several areas of different sizes at a time: a household, a neighborhood, a state, a nation. Intellectual attractions and repulsions cross time spans up to centuries as well as distances.

Most conquerors whom historians find of interest are similar in invading the space of others, separating inhabitants from their homes and possessions, and frequently having to fend off counter attacks. Where they are far enough past to acquire halos, historians tend to praise them, finding their seizing of territory impressive. Though occasionally seen differently, Alexander is typical, as is the Roman Empire in its several phases. Close up, it was Alexander the Thug, and the caesars were land pirates. All such conquerors look better at a distance than they did at the time to their victims, the most atrocious 20th century examples being the two outbreaks of militant ambition in Germany. English-speaking imperialism in Great Britain and the US looked with favor on their conquests, and still do - less so, of course, on those of France, Spain, Portugal, Holland, and Germany with help from fascist Italy. As to the last of these, hundreds of bombers and tanks and thousands of troops invading Poland were a horror everywhere except their source. Nationalism inevitably has that distortion, based foremost on proximity. In other words, groupthink attraction is to the animal kingdom what the four forces are to non-sentient bodies.

Fringes are the most fragile and frayed areas of bonds in the animal kingdom generally, and especially in people. That was no doubt behind Josiah's seeking to close outlying temples and concentrate the leadership of the Yahweh cult in Jerusalem. As T. Peter Turchin (2006) argues, a frontier is a danger zone where people team up for mutual defense. In being on their own at a distance from centralized rule, frontiersmen can also more effectively rebel. On the North American frontier progressing from east to west, the strongest resistance to advancing migrants came from veteran Indian warriors accustomed to tribal skirmishes. Overlapping or ambiguous hunting areas resembled perpetual frontiers in the skirmishes of Comanche against Kiowa, Blackfoot against Paiute, Navajo against Ute. Both sides of the North American frontier - the migratory movement and indigenous people - had the toughness and callousness in killing of chronically embattled veterans. Some of the most able commanders on the immigrant side were veterans of the Civil War, which lasted long enough to make skirmish lines there, too, like frontiers. To this day, rival patriotism keeps anti-federalist sentiment alive in the American South, where debates over Confederate statues and the Confederate flag have lasting repercussions.

Not all separatist movements come at the federal level. Proposals to partition California, for instance, have been based on the unease that rural areas feel about the domination of population centers. Where one lives is, again, what counts. Across the West, urban areas tend to be progressive and outlying areas conservative. Conservatism, in that case, means protective of local interests. As exemplified by apartheid in South Africa and segregation in the US, that us/them factor has sometimes been racist as well as regional. Proximity, to repeat, comes in degrees.

Banding together in living creatures produces coveys, clutches, gaggles, prickles, congregations, colonies, flanges, swarms, sieges, droves, armies, bevies, shoals, skulks, scourges, pods, dazzles, and other clutch names. Without a constitution and civil order that allow mobility, it is rare for a single career to carry from the perimeter of a territory into the ruling center, but now and then one does. In Herodotus, for instance, the Median leader Deioces travels from a "man of mark in his own village" (45) to the capital as a monarch whose first major command is to order "a palace worthy of a king" (46).

To command loyalty across an extended area, authority does need to be plainly visible, one characteristic of proximity being its tangible contact. The dictators of Turkmenistan put that into acres of white marble buildings and large statues, to such an exaggerated extent in the current example, Gurbanguly Berdymukhamedow, that outside the country he has become an object of ridicule. Even more exaggerated were the oratorical blitzes of Hitler and Mussolini in World War II. In comparable situations, royal progresses, military displays, dress, art, and architecture serve a similar purpose in pushing a visible ruling presence out to scattered provinces. The larger and more segmented a national unit, the more fracture lines it is likely to develop. The rebellion of former Soviet Union countries, the Balkanization of eastern Europe, and the 2014 vote in Scotland on membership in Great Britain are recent examples, the latter a vote in favor of the status quo for the time being. Despite their patriotism and constitutions, federal unions, too – thanks to the inherent strength of local or regional bonds – tend to be less cohesive than some of their inner groups.

As usual with anything involving the symbol user, however, that has mixed levels of loyalty in both individuals and identifiable units. Few have understood that better than the 19th century social novelists Balzac, Dickens, and Hugo - writers in the midst of the industrial make-over of stratified societies, with moneyed classes mixing with clannish dynasties. Speaking of "the magnanimous vanguard of the human race," Hugo in *Les Misérables* finds the vanguard shaping into an insurrection that "shoots spies [...] executes traitors [...] suppresses living beings and flings them into unknown darkness." The revolutionary movement has to be tightly knit to change the civil order. It may or may not hold a city or geographic area, or be class-related, but it must be self-identified. If it succeeds in seizing power over a broad area, it, too, then faces opposition from more tightly knit local units. Ideology and doctrine aiming to go worldwide have that inherent bonded strength of local groups to overcome. Once a revolution or an evangelical campaign succeeds, it may become as doctrinaire as the regime it has overthrown. When it carries its program throughout a population, it is likely to find the larger body less enthusiastic than the core reformers. The hardening of dogma in revolutions has led some historians to consider leftist ones an initial stage of fascism, as 20th century versions disassembled stratified societies in the interests of an idealistic order only to veer right in the Stalinist and Maoist patterns. When a Comintern replaces a Tzar, it is as if the second phase had been hidden all along in the first; it takes another hardline order to overthrow one. The entrenched center uses spies, imprisonment, exile, and executions to strengthen its hold on its target population.

Those purges are as familiar to historians as clan, tribal, and national wars. In religious and ideological crusades where the regimes seek to become international, they must proceed either by local recruitment or by force. Emissaries from their centers of power transport the belief systems to outlying places. Throughout the Americas, European-based powers and beliefs were transplanted locally by crusades and armed force. Whether based on the economy, ideology, or religion, any new movement enlists members whose only common interest may be dissatisfaction with the status quo. When the invading or revolutionary group has to define its own program, what it enacts is likely to exclude some interests of the original coalition - for example, the Interregnum in the settlement of the Puritan rebellion under Cromwell wasn't hospitable to the Levelers and Diggers who had helped eliminate the king. While revolutionary groups discourage variances, a rainbow democracy allows individuals and groups to contest one another, under rules, within a shared territory. Provided that fundamental civil rights are protected, the lesser gaps among adjacent groups make compromises more feasible. This helps keep the cracks and fissures from widening into major fractures.

It is time to remember how objectivity and the natural continuum place each in every and every in each. Groups joined by demonstrable axioms and common truths can minimize local factors and collective fervor. Objectivity seeks universal laws, applicable locally but basically unlocated. In terms of grouping, no one splits off from or attacks those who agree on re, and no one is expected to say with hand over heart, "I pledge allegiance to $E = mc^2$ and to the principle for which it stands." That stronger rational factor, at its broadest, puts local matters in the perspective of the universal laws of matter. Objectivity is often at odds with the local and the subjective in that respect.

Refugees

One other category of social breakdown needs mentioning: parties cut loose and drifting, with no way to regroup and no geographical center. I'll reserve part of their situation for later in taking up home ground in a different setting, but large numbers of people cut loose also fall under bond breaking and disconnection. That includes outcasts, exiles, transported prisoners, refugees, and those who have simply been shipwrecked or gotten lost. The worst manifestation of breakdown, next to battlefields and disasters, is a field or train of refugees whose former social order has collapsed or been overcome. With exceptions, if the areas they enter are already occupied, they are unwelcome. The refugees of Syria recently, and of other places formerly, have reached the multi millions. Where nomads make a life of wandering, refugees have it forced on them and lack the necessary preparation and foraging skills – which, in any case, have less value with high numbers. If lowland flooding and wildfires reach the extent they may under climate change, future homelessness could make former examples seem moderate by comparison. Maps of projected flood areas under rising seas don't spare sizeable metropolitan areas – in fact, a good deal of Mississippi delta land has already gone under.

Displacements on a sizeable scale aren't unprecedented. Volcanoes, earthquakes, floods, and tsunamis have been some of their sources, as in the Santorini explosion and the Minoan or Thera eruption some 3600 years ago that destroyed a thriving island and much of the Minoan civilization. For what happens in such cases, Fred Guterl, editor of *Scientific American*, cites a number of possibilities, suggested initially by Tom Leonton. Anyone living in northern Africa before the Sahara took over "would never have been able to predict it. Things seemed to be evolving in a slow, gradual fashion, but then suddenly changed dramatically" (69). Similar radical changes have come in other extinction episodes. Not to be entirely ruled out is a planetary flip of the magnitude as the Permian-Triassic extinction, a "cautionary tale, a preview of how climate change can annihilate life on Earth" (42). A great many species are in what the world-traveling journalist Alan Weisman (2013) calls a countdown, endangered not by some particular episode of magnitude but by gradual habitat erosion and incremental climate change. The ice ages brought such changes on a global scale.

Losing place not only politically and socially, but in the cosmos is nothing new in eschatology. This has been a recurrent philosophic and religious topic under the illusion that major natural events are intentional, an odd outcome of which is limbo, where contrary winds blow homeless creatures about, broken loose from regional, social, political, ideological, and congregational connections. Milton's version of such a place is a curious neutral zone between the regimes of God and Satan, containing remnants of the paradise of fools and the addle witted, neither consciously evil nor virtuous. Ariosto's debunking of theology (Canto 34 of *Orlando Furioso* or *The Frenzy of Orlando*) made a noteworthy contribution to that literature of scrambled minds and flying objects, a literature of human debris and disconnection. As such a place illustrates, the progress of a population, even of refugees, toward something is itself a gathering force. Lacking a mutual goal scatters a population, as if members of a wagon train headed off in different directions.

For symbolic purposes in his 'anatomy of the world', Donne uses the death of Elizabeth Drury as an occasion to go as far in such scattering as anyone can short of an apocalypse. When coherence leaves, not only the planet but its "immaterial parts" feel a "consuming wound," and beyond that the heavens, too, fall apart:

> We think the heavens enjoy their Sphericall Their round proportion embracing all.
> But yet their various and perplexed course, Observ'd in divers ages doth enforce

> Men to finde out so many'Eccentrique parts, Such diverse downe-right lines,
> such overthwarts, As disproportion that pure forme. It tears
> The Firmament in eight and fortie sheeres. (*The First Anniversary*, 251-258)

Similar prospects of coming unglued have surfaced in periods of stress as they have in modern times in post-apocalyptic fiction and cinema, the end in modern versions usually coming as a result of a nuclear holocaust rather than a natural disaster. Yeats, in a time of war, found a remedy of sorts in fascism, as did Ezra Pound in World War II. The 'bundling' of fascism is a common response when cohesive strength gives way to agitation and brings the huddling of the fearful. Limbo in mythic terms, refugees and outcasts in human terms, and rubble in material terms warrant a more prominent place in common opinion and in philosophy than they are usually given.

Any breakdown phase necessarily finds bonding forces giving way to individual unit vector and momentum, if not to the extent of eschatology and literary limbos. Large size rocks, cliffs, and mountains are the broken remnants of something once molten and cohesive. Bits and pieces up and down the scalar spectrum tell that story even as they testify to something previously bonded.

That anything now separated had to be first joined is as fundamental to atomic matter as hydrogen feeding into the elements of the periodic table. Free floating hydrogen that doesn't undergo that transformation is in a sense in limbo. Though intact in themselves, the atoms aren't bound in higher level composites and may continue free forever. Celestial examples of disintegration in nebulas extend in light year distances. None of the binding forces reaches far enough, powerfully enough, to configure the entire Great Welter in its current expansion mode. Its parts are drawing ever farther apart at high speed, not faster than the speed of light, one section relative to another, but fast enough that a bright object sending out light 13.7 billion years ago could now be several times that far off. That is disconnection at high velocity in high volume indeed. If we ask what holds the universe together, the apparent answer in this phase is nothing. Gravity isn't currently doing so, and no other binding force is capable of extending that far with sufficient binding force. That the dispersal might eventually reverse hasn't been ruled out. Should a big crunch come about, that too would be beyond imagining.

CHAPTER NINE

BUNDLING THE UNIVERSE

The Argument

Philosophy and science are comfortable with lumpsum terms in placing one thing in relation to another and in classifying categories. In representing much in little, these are similar to sampling, foreshortening, and fast forwarding. Precise measurements have a place as well, but obviously more in the microcosm than in the macrocosm.

In a Nutshell

Like cartoon depictions of evolution from fish to quadrupeds to *Homo sapiens*, models, diagrams, and equations offer panoramic overviews. When these are granted the liberties they often take with nature's diversity, it isn't as strange as it sounds to package its categories and all of it in total, the latter in the *visible universe,* 95% of which is not visible. *Uni* takes further liberties, this time only hypothetically with a possible multiverse that could have arrived with the one we partly see, or at some other time. Coming out of nowhere in that context means what *Nothing*, or what ancient seers sometimes called the void, does when put beside *Something*.

Despite these quite serviceable means of citing the natural continuum as a whole, I find the discourse, the lumpsum mathematics, and the visual representations in use more troublesome than might seem justified. They have, after all, worked well enough since advances in magnification brought more into view - small, large, and distant. The main problem – back to the premise at work throughout – is that our normal sensory range doesn't prepare us to take in extremes in size and movement such as quanta, galaxy clusters, and light year distances. The coarse grain measurements and analogies can't deliver what we aren't prepared to receive, just the building block level alone began to drive particle physicists all but mad in the 1920s. That wasn't primarily a discourse problem, but it involved one in its attempts to transmit, in some intelligible way, what shouldn't be but apparently is. In the interim between then and now, some of the puzzles and paradoxes have become familiar with repetition, but they remain badgering. It isn't only laymen who find it difficult to cope with the stretched scales, but scientists and mathematicians as well. A teacher of automobile mechanics and a master carpenter have something at hand to point to; no epitomes or samples are available for milliseconds, light years, megaparsecs, antineutrinos, or even electrons.

The nomenclature of the particle zoo and accompanying descriptions cited earlier, instead of getting clarified and simplified in the early days of quantum mechanics, became an ever greater entanglement. Some of the puzzles arose from the distinction between modern physics and Newtonian or classical physics, which Sunny Auyang (1995) locates mainly in spacetime relations: "Going from classical mechanics to quantum field theory, the focus of physics changes from locomotion to dynamical interaction.

The primary dynamical concepts change from action-at-a-distance [Newton] to coupling-on-the-spot" (119). That is more or less equivalent to complicating things linear with things nonlinear, crossing, doubling or circling back. Physics and astronomy are typical in the gaps they open between common language and what the scalar spectrum calls for. The kind of field used in *quantum field,* for instance, bears little resemblance to one raising rice in Japan.

In scientific use, *field* designates something uniform in its sections, and extended rather than concentrated. A gravitational field, however, if we are to include some of its quirks, would be assigned hills and hollows, since gravity alters under extremely condensed black hole conditions. Although the strength of a gravitational attraction does always increase with mass and diminish, as Newton specified, with distance, comprehensive rules for it weren't proposed until Einstein's general relativity (1916), and in that context the 'field' was typically pictured as a webwork sagging under a weight, bending spacetime according to strength and distance.

Some of the apparent puzzles that put kinks in explanations, such as what happens when a beam of light shoots forward inside a high speed space ship, clear up upon examination. That should logically break the speed of light law, but doesn't. The spaceship might as well be parked on the moon. We've grown used to that, and accustomed to astronauts traveling through space differing in time span from the apparent same moments for those tracking their progress in Houston. It is also common knowledge now that, by earth standards, clocks run slower and measuring rods shrink when they move at high speeds.

Such things no longer puzzle those who have been exposed to them, but how an electron can be both a particle and a blurred field, and present in more than one place at a time, remains mischievous. The extra dimensions of string theory present a different set of problems when expressed in two dimensional terms. Saying that the extra dimensions are 'wrapped up' in the strings doesn't help a great deal. Leonard Susskind's and George Hrabovsky's *The Theoretical Minimum: What* You *Need to Know To Start Doing Physics* (2013) cites comparable examples. Strings, membranes, waves, fields, and latticework vary the metaphors and diagrams, as do Richard Feynman's ingenious charts of colliding electrons and photons. In the latter, the particles can destroy one another and produce virtual equivalents. Adding antiquarks and anti-versions of other particles shows why diagrams use not only straight lines, arrows, equations, and squiggles, but also explanatory footnotes.

Problems in representation aren't limited to science and the micro and macro extensions of atomic matter; misreading something and confusion are obviously nothing new, merely more formidable when they take in a great deal of the natural continuum. Since the invention of symbols and probably in the preceding proto language era of gestures, scratches in the dirt, and facial expressions, some signs would have been puzzling. As freshmen and sophomores find, literature too thrives on ambiguity. Very little comes straightforward and unambiguous in Shakespeare. We band of brothers, we happy few in *Henry V,* genuine patriotism or rhetoric? In that and much else, we should probably go with the complexities modernists like to expose in anything high sounding. Ambivalence and ambiguity crowd into bed with it. A more comprehensive barrier between common understanding and the stretched continuum, top to bottom, is that while all parts of it are in motion, means of representation usually freeze a given moment. Even in moving pictures, movement is an artifice. Successive frames can show the planet approached from outer space as a distant blue dot, then move in a few seconds down to outlined continents, and in a few more land in Times Square. The order of magnitude in distance drops in a few seconds from about 10^6 miles to a few kilometers. Within a timeframe more extended than usual, a mockup of drifting continents in video form could depict travelers going from North America to Europe without swimming or using anything that floats or flies.

Such devices aren't necessarily misleading, but they point up the gap between almost any form of representation and what it seeks to represent. Back to assigning a name to the *visible universe*; it suggests more insight into it than anyone actually possesses. The label has aided and abetted numerous misconceptions, including, as we've seen, lining up high powers behind nations. The name and a master controller give assurances about design and control that, given what the universe contains and the vacancy into which the galaxies are scattering, are unwarranted. What bunched dust and gas into stars and sent galaxies on their way most likely wasn't Aten (Egypt), Yahweh (Israel), Ojebwe (American shape shifter), or any of dozens of other human-like figures put in charge. These don't necessarily come attached to *universe*, but that they often have is due, in part, to the multiple applications the term allows when it breaks loose from the history of atomic matter.

Collecting Samples

Whether as small as dust motes or as large as galaxies, specimens and samples aren't abstractions but material examples of recurrence. They are different from epitomes and obviously quite different from models. Whereas the latter are inventions based on core properties, a specimen starts with what is there and is singular in location. Even when locked together in rock, specimens and samples aren't necessarily uniform. Breccia, for instance, differs from one sample to another. Yet any given specimen typifies that particular mixed kind of rock. Whilst in igneous and metamorphic rock, the formation sequence works the same every time in producing quartz or gneiss, sedimentary rock ranges from relatively uniform gypsum to the broken-piece mixtures of breccia. The samples, in other words, are less narrowly exact than in kinds that tell of more exactly programmed formation sequences. Suppose, for instance, that astrophysicists were to investigate the large asteroid Vesta, or a mini planet such as Ceres, and apply the crystals they find there to earth samples. If the samples turn out to match, the logical conclusion will be that space object also once had magma. The sample was the product of a known process. The next step of such a research team, a real one in this case led by Beverley J. Tkalcec of Goethe University in Frankfurt, is to publish its findings and explain how it has reached its conclusions. Going more broadly public might require a science writer such as Amina Khan presenting the findings for a newspaper such as the *Los Angeles Times* (21 January 2013). By stages, the far off and the complex then get reduced to something more manageable.

For uniformity in small sizes within the visible range, consider common dust again.

Since it doesn't show what its former composites were, it isn't often used as a sample of anything, but it is nonetheless something of a marvel. Both the planet's atmosphere and space are full of it. As a small but visible level of material breakdown, it gestures toward even smaller levels out our sight range. Like gas molecules, dust motes lack viscosity and don't conduct heat, and yet do sometimes form into visible clouds. On the planet, when they settle and turn into compacted dirt or with water into mud, they can clog a machine or bury a civilization. Clouds of them can thin and thicken and move at a measurable speed, but just that and nothing more doesn't tell us much about what its specks actually are. The settling of denser clouds in space shows gravity at work and, under known celestial mechanics, astronomers can deduce the further collapse of whirling disks in preparation for star formation. It is common to say that humans, too, are made of star dust, and, certainly in more than one way, we are dust collectors.

Sampling runs into trouble when the phenomena are in motion as dust usually is. The problems for representation and dissemination that come with that haven't attracted as much attention in discussions of discourse or been put to work in critiques of the history of ideas as one might expect. I cited some of that commentary on discourse earlier. Others examples include *Discourse Studies* (2003, 265-279) in which Grey Meyers cites still more as do Harvey B. Sarles (1985), Annamaria Carusi and Aud Sissel (2014), Lorraine Daston, Michael Lynch, and Steve Woolgar, (2014), Norman Fairclough (1992), Leah Ceccarelli (2001), Ray Jackendoff (2002), Merlin Donald (1991 and 2001), and Carmen Pérez- Llantada (2012). The fuller extent of the lumpsum method when we put words and numbers alongside atomic matter samples is revealed in Amir Aczel (1999), John D. Barrow (in numerous books, 1986, 1991, 2000, 2008), Robert Oerter (2006), and Lee Smolin (2001). In contrast to these, commentaries on discourse that drew largely on the Greco/Roman tradition didn't normally venture into extremes or into depictions of chaos, limbo, and miscellany. Nor have natural philosophy and science typically done so specifically as problems in discourse, the basic one besides incessant motion, to repeat, being that our resources developed within a narrow range.

Where the problem lies less in extremes than in the profusion of objects, chaos theory's advances extend the means of representation. Steven Strogatz lists functional genomics, synthetic biology, and econophysics as collaborative fields open to limited confusion, especially self similar bifurcation, which multiplies its instances in smaller and smaller sizes. Of special interest over the past few decades have been Mandelbrot sets and fractal geometry that take a set form and reduce or expand it by multiples of ten. Some visual demonstrations have drawn on Moebius strips and Escher stairwells going up and down in the same direction, thus:

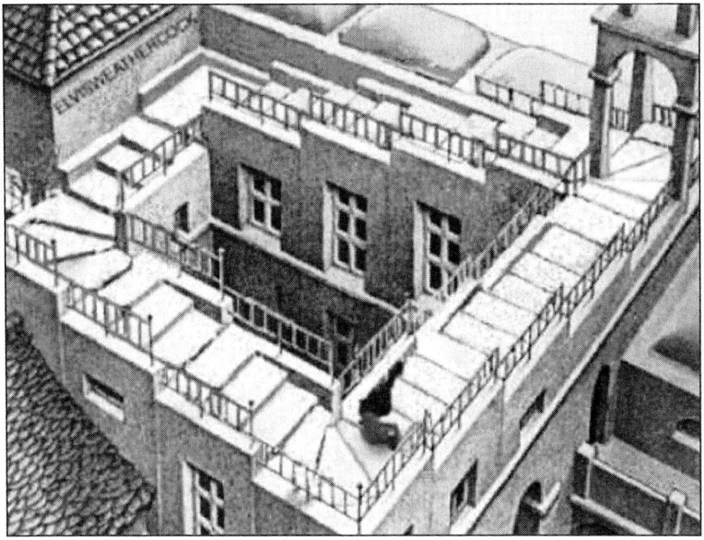

In Barrow's succession of "key images in the history of science" (2008), one of the more striking combinations of manmade symmetry juxtaposed with nature's is Escher's "Moebius Strip II' (323), which has red ants crawling on a twisted machine-tooled strip:

Ants have a uniformity and symmetry of their own, but putting them on a construction of squares twisted into a chain underscores the fact that manmade mechanisms are more exact geometrically and, of course, aren't alive and crawling. Behind the ants and the mechanism, in this case, is the artist Escher making a mental game of comparison-and- contrast and inventing bizarre tricks of perspective. The examples are representative, but not of something straightforward, rather of paradoxical quirks. They are in accord with quantum mechanics in seeming enigmatic.

The Escher tricks look bizarre until we stop to think. In the framework of the continuum and the scalar spectrum, it may not be sprawling and prolific atomic matter that is disproportionate and bizarre, but our limited range and reliance on common sense. Some such double take shows in other fields as well, in drama, for instance, when it adds a metadramatic level. Shakespeare is especially noted for plays within plays, and for being self-conscious about off-stage play acting. Tom Stoppard's *Rosencrantz and Guildenstern* compounds that with a play about a play enacted as if before a fun-house mirror. A sizeable share of modernism has been characterized by awareness of a given medium's disconnect from plain reality. Much of Stevens' opus concerns poetics in that mode, in poems about poems and in the long philosophical poem "Notes toward a Supreme Fiction." The expansiveness of *supreme* in the company of the modest *notes toward* is equivalent to making a stairway's up look down and its down look up.

Variously Scaled Models and Cascading Files

The foreshortening devices of language that make adjustments through samples, abstractions, and epitomes are effective only where what is selected is truly representative. Indicating the scale in use is usually a necessity, but a single sentence can put quite different levels in play. Saying "mice have ears and tails" links one species to others in the ear and tail branches of the animal kingdom and spans various lengths of time. In pinning them to elephants, lions, and horses, we are on the trail of an evolutionary succession in which ears and tails find a common heritage. Any sample indirectly refers to that origin- in-stages, perpetuated and multiplied in many species. The small tip of a residual tail in humans, together with ears, tells us something about the ancestry we share with quadrupeds. How the ears of elephants got large and tails small, and just the opposite in horses, might not be possible to trace with certainty and not very rewarding, but the natural succession tells us that environment governed throughout, as it always does. Presumably nothing gets perpetuated unless it served a purpose over a long span of time. When we add reproduction to survival as a reason for something's existence, showy feathers and other features of conspicuous attraction claim a place among a species' other advantages. Such aspects of biological continuity make sampling work reasonably well. Scalar measurements figure in the timing of evolutionary stages, and gradations of one kind or another.

As we've seen, cascading categories from general to specific are the most common devices with which to organize nature's vast production. The various levels of abstraction aren't necessarily meaningful in themselves, but they help separate what counts more from what counts less, with more general terms rising in conceptual and organizing prowess.

Captions without content are empty; content without captions is chaotic. In variations, that has been a working principle for sorting things out empirically - as it was in Einstein's remark, "science without religion is lame, religion without science is blind." (*Religion* is sometimes interpreted as *philosophy* on the grounds that Einstein intended no reference to personifications.) Generalizing on the basis of samples requires rising on the taxonomy ladder from species, to genus, to kingdom. That common cascading file system is itself something of a specimen, an example of how science and philosophy go about organizing their materials. The master concept that catches everything is what I've been calling a term of terms, the *natural continuum.* Philosophy and religion have traditionally substituted something like a *summum bonum* or *god* for the total, but that invites the problem mentioned earlier of a nature becoming intentional, thereby converting something harsh into something malignant. This is easily enough avoided by not resorting to personifications; we've no compelling reason to moralize the seas, atmosphere, stars, or lifeforms. That many of the latter need protein to survive, returns us to Blake's tiger/lamb problem.

If we want synonyms for the continuum on a ladder of inclusive categories, we have other abstractions to choose among besides *universe* - *Something* as the opposite of *Nothing*, for instance. The sub-captions down to species and to specific instances get quite numerous. In that respect, categories enable a cleanup operation much as labeled storage bins police what in a loose state is clutter. In the animal and vegetable kingdoms, scavenging converts some disintegrated and scattered waste into individual property, one of nature's own ways to clean up and reorganize. It has ways to do so in the non-sentient realm as well. What is flying loose is rubble, until it orbits something or lands somewhere and becomes part of an integral body, and that is true all the way up from neutrons to galaxies. We know approximately how much has come together in nomenclature like *the solar system*, which earns that title by moving as a whole relative to other stellar systems. The chemistry of mixes sometimes makes a particular ingredient more important, and can also make a composite more than the sum of its parts, in a similar way to yeast rearranging other ingredients of bread dough.

As we saw in the chapter on patterns, *order* in all categories of pattern making and classifying is a convenient term for a schematic arrangement of parts, stable enough to be identified and labeled. Fields and other area designations offer a more panoramic level of organization, usually an area without subdivisions, though grids can be superimposed on them. Gravity, as one such field, extends everywhere matter and space do, and generates directional crisscrossing among multiple attracting masses. That illustrates how a singular law or set of laws can produce a hodgepodge as well as a functional congress of parts. For gravity specifically, putting the influences tugging on planet earth in order isn't easy - what with Mars, the moon, and the sun contributing, and these plus Mercury, Venus, and Jupiter in motion. That is more a measurement than a naming or conceptual problem, and it doesn't undermine the Newtonian inverse ratio law. It merely shows why representations don't attempt to account for everything, and why common language as well as science is usually coarse grain in what it designates. Multiple weak and distant gravitational attractions aren't allowed to complicate tide tables based on a three body influence.

Salience, with an assist from proximity, says that it is mostly what is near and powerful that counts. On that principle in human affairs, any given nation's heritage is more important locally than it is elsewhere. What goes in Shanghai doesn't in Chicago.

How much data is now available to fill out the models, charts, and generalized schemes? That's impossible to say, but as I've suggested the archives and repositories are overflowing, and according to the website "how stuff works" computer assisted stuff may one day extend to Cotta quantities, or 2^{80} bytes. The human mental apparatus may come near such quantities in receiving impressions every waking hour of every day plus dream states, but consciousness remains unaware of most weak and passing impressions and so isn't overpowered by them. The brain's main sorting occurs before consciousness takes over and assigns names. A few impressions spend a limited time in memory's leaky storage. If the remaining bits and pieces could be assembled, they would be an unstructured heap bearing no resemblance to a filing system. Listing millions of books in an indexed repository puts comparable library quantities in card catalogues. While none of this denies the necessity of sorting out the files and categories, philosophically speaking we do need to give the disintegration phase of nature's cycles and advanced stages of dividing and multiplying their proper due. Most worshipful concepts of cosmic powers fail to do so, however. The breakdown and separation principle is like the crisscrossing gravitational tugs on the planet in that respect; the components of integral objects and systems sooner or later separate. The reason for heaps of facts to begin with, to underline the obvious, is the prodigal amounts nature turns out in its dividing and multiplying. The number of large spheres alone is mind boggling, and again, remember, none of them is perfectly organized or geometric. Like the boiling and spitting sun and the cracked, volcanic earth, they have their perturbations and their term limits.

Next to abstractions and cascading files, math is the most common way to compress much into little, as the figure10^{-21}s is an effective way to cite the duration of a particle. The billions of collisions and readouts produced at CERN get reduced in the handling to a manageable number. The 'excessive' numbers in Pythagoras have sometimes been thought to be built into the order of things, but have most likely never been anything more than steps carved up a hill. The system is imposed, not built in. That the integer divisors of 12 (1, 2, 3, 4, 6) add up to 16 means nothing. The sum of divisors in 6 (1, 2, 3) make that number, which is sometimes called perfect, again without any corresponding perfection in nature. Outside of spheres, most material solids don't approach anything geometric, any more than clouds do that come in billowing, flat, and towering shapes. Nature doesn't know numbers or the names of species - it doesn't so much as add 2 and 2 to get 4.

As distinct from samples, invaluable in their compression, are mockups and models.

Silvan Schweber (1997) and Robert Oerter (2006) value the standard model above any other, It "will be remembered – together with general relativity, quantum mechanics, and the unraveling of the genetic code – as one of the outstanding intellectual advances of the twentieth century," Schweber believes (645). This, in my opinion, is true of possibly any time, not just the 20th century. Charts and graphs are like generalizations in their capacity to pull back from particulars. The following one (from the US Department of Energy and the Contemporary Physics Education Project, CPEP) is another rendering of standard model elements, more complete and more complicated than the version displayed earlier:

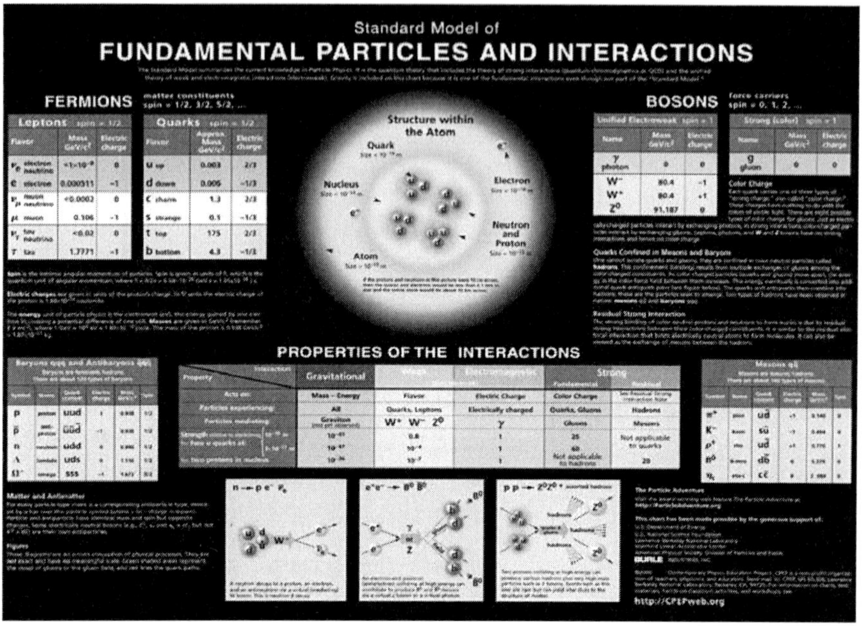

As this illustrates, the natural desire for simplicity, nearly everyone's preference, has produced the opposite. The basic laws aren't themselves that overwhelming in number, but what they produce is, both in instances and sub categories. If a Theory of Everything (TOE) ever does get formulated it, too, won't be simple. Again, the standard anthropomorphic way to simplify is to propose an all-knowing figure that sets aside much of what exists - either that, or something incredible in the mythic line such as that the volcano that just erupted in the Hawaiian Islands derived from a momentary lapse in the behavior of the first humans.

Randomness

Where atomic matter is ruled by necessity, randomness, strictly speaking, doesn't exist. Determinism, from the Greeks to Pierre-Simon LaPlace to the present forbids it, and I don't see any way around that. Atoms don't have minds of their own, and we've seen the drawbacks of having a higher power resembling mankind do the manipulating. With the exception of intelligence and free will making selections within a narrow range of options, natural laws without fail determine what comes next. The proliferation, velocity, collisions, and crossing vectors of composites are what generate the confusion. With sharper and more distinct imaging, it has become clearer that more of natural history does, in fact, look messier than traditional anthropomorphic variants supposed. Concerning apparent randomness, philosophically speaking we need only acknowledge it in general without itemizing it. This is as true of what is visible within the human range as it is of the rest of the scalar spectrum. Poets, philosophers, geologists, and nature writers have different approaches to that weed-patch confusion, but they all rely on salience, proximity, and cascading tiers to do the cleanup and organizing.

As Milton found, giving a supreme power a humanized form and making it omniscient required adding Satan and his wrecking crew to explain roughhewn nature. Even then, Milton didn't want to grant sufficient power to a former archangel to raise mountains or freeze and overheat climates, and so fell back on the *omni* power to perform the actual wreckage. Neither he nor any of the astronomers of the day knew about the celestial levels of that wreckage. Turning that, too, over to a supreme power created an additional problem with God's ill-tempered wrath. Gods and goddesses earlier than the Hebrew source for Milton's creation likewise had a tendency to get angry, quarrel, and wage wars, but Milton's version should logically control his temper enough concerning the planet not to kill everything that lives. The greater confusion that has come into view with telescopes makes it even more illogical to attribute detritus and wreckage to angel and human misbehavior. With the breakdown phase of atomic matter, no one needs Beelzebub.

Buildup and breakdown under the four forces accounts for everything both misshapen and symmetrically patterned. That doesn't interfere with free will, which finds its options within the small range available to biology.

Marvell's country house poem "Upon Appleton House" (1651), written in the wake of the English Civil war and a century and a half before the French Revolution, chooses one of the prominent options of his day (and ours) in concluding that the local topography came to be what it is by a random sequence. Without any notable influence from outside nature, sections of it look like

> a rude heap together hurled,
> All negligently overthrown, Gulfs, deserts, precipices, stone.
> ("Upon Appleton House," 762-764)

Negligently means without visible design. All an inventory can do is collect the wreckage under general class names like *gulfs, deserts, precipices, stone*. The samples could be extended to cliffs, frozen wastelands, jungles, the fury of storms, and so on, but the general title, "a rude heap," tells us all we need to know. Had telescopes been more readily available, Marvell might have carried that principle into space and multiplied the examples. As a friend and younger associate of Milton in Cromwell's cabinet, he might also have conceded a population of devils somewhere about, but, if so, he never explicitly assigned the planet's roughness to them nor to divine wrath in response to them.

Thanks to microscopes and telescopes, both still novelties in Marvell's day, making the small large and the large small brings another factor into play in scenes of mixed order and disorder: point of view. Where an observer stands influences the perception of what are cattle from one distance and from other distances facial blemishes, fleas, and stars in constellations. They

> shrunk in the huge Pasture show As Spots, so shap'd, on Faces do.
> Such Fleas, ere they approach the Eye, In Multiplying Glasses lye.
> They feed so wide, so slowly move, As *Constellations* do above. (457-464)

This scale-playing reminds us that the proliferation of natural forms is only part of the bundling problem. The angle of vision combines with it in multiplying the possibilities. As special relativity has added, different velocities indicate what travels together in integral objects. The cattle are small compared to the pasture and move slowly, just as from, the planet's distance, constellations appear to do. That seven stars are grouped in the outline of a big dipper is disconfirmed from every other angle, as is the hunter Orion confronting the bull Taurus. Counting the big dipper stars always produces seven, but stopping with just them is arbitrary. What the stars in total have in common with grazing herds is their scatter, except without the comparative uniformity that level or sloped ground gives a herd. Another factor in singling out and arranging star groups is their luminosity, a salience feature in which what is brighter is more important. Salience says, in effect, that what counts here and now is the shape of the dipper outlined by exceptionally bright stars. Here and now, in this place at this time, those particular stars do resemble a tin cup with a long handle. That we are being imaginative in saying that is clear to everyone, as it is again in seeing the shape of horse's head or a swan in a nebula. We know that is more imagination than observation, and mainly a way to refer others to a particular patch of the night sky.

The mixed velocities, sizes, and crossing vectors, plus point of view, have built-in confusion, but the basic laws behind them are known and can be found threading through them all. In naturalistic terms, we know that earth's plate tectonics, molten plumes, vents, volcanoes, subduction, and earthquakes – these in various combinations – have roughed up earth's outer shell and piled up the rude heaps. That can be explained in terms of how atomic matter behaves when gases, solids, liquids, dust, and viscous substances of different temperatures come in contact, moving in different directions at different velocities. As we've been seeing and as this indicates, it isn't only extremes of measurement or combinations of mass and velocity too slow or too fast that challenge our means of representation, but objects too numerous and in disarray. Even without that built-in confusion, Robert Pollack wonders if, in being the product of evolution, the brain hasn't become outmoded - especially in the age of AI: "Just as the past has established that there are odors we cannot smell and colors we cannot see, there may be ideas we cannot consciously articulate" (*The Missing Moment*, 36). We don't take naturally to the thermonuclear invariables behind heat plumes from earth's core to the surface, or to a roaming Jupiter long ago altering the orbits of other planets.

Rutgers Bergman, in *Utopia for Realists*, finds computers overtaking brains in making calculations, and nothing prevents them from becoming a billion times smarter than people are. The continuum already tests our limited brain power, as, for instance, in the proposed period for the inflationary universe that Alan Guth makes "responsible for the creation of essentially all the matter and energy of the universe" (15). That heated inflation might have lasted only 10^{-30} second, during which "the universe expanded by at least a factor of 10^{25}, and perhaps a great deal more" (14). Despite knowing what a second is, we can't comprehend anything that fast or powerful enough to affect the entire universe, even compressed as it was at the time. As Lee Smolin estimates in *Three Roads to Quantum Gravity* (2001) that the quickest thing any living person can snatch out of the air "takes more than 10^{40} fundamental moments," the minimal Planck time units (5.39×10^{-44}s). The "blink of an eye has more [of these] than there are atoms in Mount Everest" (62). Try putting *that* in familiar terms - or try imagining the entire universe compressed into something smaller than a pebble, which cosmologists do for one of the most inconceivable of natural history phenomena. Nano technology and the burgeoning sciences of the small and fast operate in measurements almost that inconceivable. Only computers can handle them, and even they aren't good with miscellaneous heaps, requiring regularity in placement, velocity, size, and whatever else can be named, pictured, categorized, and quantified.

Nature writing sometimes assumes some of the responsibility that science and philosophy share in accounting for nature's roughness and scalar range. Try putting a John McPhee passage on geology beside the prose of a Greek or Roman rhetorician, or a stylist in English such as John Donne, Joseph Addison, or Richard Steele. How does *Assembling California*

(1993), and the company it keeps in *Annals of the Former World* (1998), fare beside the verbal hurricanes of Donne and Lancelot Andrewes, based as these are on a divinely chartered, human-oriented universe? Surely quite well on a match-up-with-reality test, although reading prose responsive to large chunks of factual material can be like watching a wobbly, stiff-legged foal learn to walk:

> A few tens of metres of ocean sediments drift down upon the deep cold slab, settling on top of
> a kilometre or so of pillow lavas, under which is a kilometre or so of sheeted dikes, under which is
> a kilometer or so of plutonic rock (plagiogranite, gabbro), under which is
> a kilometer or so of plutonic rock in which cumulate crystals settled in layers upon a distinct chamber bottom–
> the Moho–
> under which is a kilometer or so of mantle rock, some of which was melted in the spreading center and some of which is peridotite in its solid original form and if water has reached it is serpentine (McPhee,115).

This passage, and others like it in nature writing since about the mid-19th century, move to trochees in rhythmic batches joined like legato in music. Maybe, in this case, that is less a stumbling foal than a lumbering Sasquatch, but some vocabulary like *plagiogranite* and *gabbro* is unavoidable once the goal is factual accountability. Rounded periods and emotion have to be sacrificed to the unadjusted or less adjusted factual stuff. In any case, elegance isn't the strong point of "peridotite in its solid form." Fuller accounts of geologic California come even more loaded with technical terms citing rocks, minerals, ages, and depths, all these most likely parallel to numerous other rock satellites.

Another who frequently turns topographical irregularities into commendable prose with extended implications is Wallace Stegner. On the prairies of Canada, eastern Montana, and the Dakotas, the land's surface approaches one of the fundamental geometric shapes, the flat plain: "These prairies are quiescent, close to static; looked at for any length of time, they begin to impose their awful perfection on the observer's mind. Eternity is a peneplain" (*Wolf Willow*, 7). Since, like the other branches of mathematics, geometry is based on theorems, a single peneplain (almost a flat plain) represents flat surfaces everywhere, curved in spacetime and this one curved on the planetary sphere but looking flat. Stegner's point is that stretching our minds across a vast flat surface is like trying to encompass eternity. Bringing imagination and reality together in discursive ways is also the practice of nature poets and, in pictorial terms, of outdoor painters like J. M. W. Turner (1775–1851) and Albert Bierstadt (1830–1902). We are familiar with that degree of irregularity and often find it beautiful, whether or not a given example has the towering vertical cliffs of a Bierstadt painting or the forever-seeming flat of the plains.

The goal is to put half organized detail in a context that enables us to understand it in comparative terms. The technology of modern terrain imaging and mapping derives from GSI (Geographic Information Systems) and from Google Earth, the zooming capacity of which carries down into canyons and across plains. Cars equipped with cameras drive the roads and streets, recording scenes for Google and Apple to splice into video systems. The visual spread of the spherical satellite serves the function of an upper level organizing abstraction. It is now possible in electronic map zooming to get overhead pictures of a neighborhood and street level shots of individual buildings. Just keep going from continents down to regions, dialing in streets and the house at 1413 Wilson Avenue. To represent the planet's composition ever more accurately like that, disciplines as different as biology and oceanography have added software to the Keystone digital earth that Google purchased in 2004. No further technical barriers of significance keep anyone from voyaging almost anywhere in virtual form and adding examples to category titles.

Simulated movement through a field of molecules has become as presentable as space travel through stars. Choose your galaxy: a space traveler can hitch a simulated ride there. Voice-over may, meanwhile, provide thematic guidance and turn specimens into examples and epitomes. Doing so comprehensibly requires initially establishing the constants responsible for them, meaning the relations of cohesive mass to heat and movement. Still, when all that is said, we have the unimaginable small, fast, large, distant, and heated extremes - plus the invisible 95%.

Modularity

Adjusting to scalar extremes doesn't differ greatly from one normal brain to another, the principle behind my snapshots of McLaurin's uneven progress. Cognitive scientists and evolutionary biologists apply the term *modular* to inborn mechanisms that don't require conscious effort or memorization - which anything unfamiliar in the scalar spectrum, on the other hand, does. None of our scalar adjustments could really be called modular, but we do have another category related to them to draw upon, namely skills learned and practiced until they become habitual and no longer require much conscious effort. Some of these may seem semi-modular, such as adjusting to objects far off looking smaller than they do up close. The main point of getting more accustomed to *V* numbers, tables, and charts is to make them more semi-modular in that sense, more habitual - also to keep unorganized wreckage more in sight as a correction of some traditions. As it is, with standard issue equipment no one understands what a light year is in the same way everyone understands the quarters of a basketball game.

That a good deal of repeated behavior can be considered almost entirely instinctive David J. Buller (2005) and others classify under the "massive modularity thesis" (126), in which brain modules were long ago programmed to develop on schedule. The 'massive' part of that assumes that they do so in more areas than one might initially suppose. (Buller cites Noam Chomsky, John Toby, Led Commies, Steven Pinker, and others.) This has undoubtedly come about by the usual means of long-term adaptability and recurrent occasions that have built patterns into our systems over a long enough span to make some operations inheritable. As it is with just the customary modularity, one section of the brain operates the heart and another eases the way into learning grammar. That some of the latter is as basic as a subject-verb-object sentence (agents acting upon objects), makes it a candidate for modularity, or what Chomsky called generative grammar.

Another way of looking at recurrent behavior is to assume a flexible, generalized brain that adapts to situations as needed without a great deal of preparation. Other possibilities besides acquired habits lie somewhere in between, such as a latent capacity that doesn't unfold until needed, and then does so with less effort than something that our ancestry didn't call on as often. The desire to explore and discover the pleasure of doing so can probably be included in that without stretching too much, since nearly every normal person has such a desire. Not the information is particular, but the gratification of reaching out, seizing something, and understanding it taps into a psychological universal as fundamental as eyes tracking movements and infants grasping objects. Merlin Donald settles for calling the human brain a hybrid, with much given and much acquired. No one would disagree with that. Consciousness isn't uniquely human, but it obviously has more to do in minds equipped with a large repertoire of symbols than it does in animals with a limited listening vocabulary and few responses in sound and body language. As Donald summarizes, consciousness changes "the rules of the cognitive game because it is a self- regulatory system, fundamental to all voluntary mental operations" (8). Equipped with tools like words, numbers, and diagrams, it has acquired a good many rules and procedures for handling nature's assortment we encounter regularly.

In combing through the debate surrounding the categories of genotype (genetic makeup) and phenotype (individual characteristic), Buller concludes that to merely say that people are both modular and adaptive is to say very little; introspection doesn't detect genetic modules at work, so not much can be determined by using it. Dennett, in *Consciousness Explained* (1991), finds the architecture of the mind lacking a headquarters or Central Meaner (253), however, that doesn't put everything on automatic pilot. Together, the brain and mind constantly combine the inborn, the individually learned, and the culturally indoctrinated. In Bennett's evaluation, the cultural models come in "thousands of memes, mostly borne by language" that "take up residence in an individual brain, shaping its tendencies and thereby turning it into mind" (254). That may be about all we can do to distinguish mind from brain in the company of sensory experience and repeated verbal formulas. Memes don't address scalar extremes; none of the verbal formulas from the past and none of the canned and stored brain modules gets us accustomed to them. Even within the sensory range, nearly everything passes us by. Consciousness translates some incoming sensations into conventional symbols, but leaves most of them unformulated. The memes it applies to them come pre-packaged and ready for application, but they don't apply to most of the visible universe. Partly, what practicing the scales is meant to accomplish is the *extension* of the practiced and the habitual, and thus the eventual incorporation of more of the scalar spectrum into conventional terminology and consciousness. As it is, no meme says anything about 10^{-43}s.

Because social formulations within the sensory range have to be sorted through and applied to particular episodes, they aren't modular, only conventional – that is, cultures add to them over time. Education, formal and informal, transmits them. The rhetorical devices in the manuals are repeated often enough to become commonplaces. Every native language user acquires a large store of them and of idiomatic expressions.

Exclamations, questions, and declarations are modular in format, but particular in application. Modularity presents something of a conundrum in that respect, in that massive modularity holds that whatever people think they are inventing or choosing, an inborn module or acquired memes – a cultural implant – has put there. Pursuing that line too far negates moral judgments based on distinguishing choice from necessity. It converts first degree murder into manslaughter and theft into circumstance. In assuming that free will has greater latitude than that moral philosophy takes up where natural philosophy leaves off. The extended prehistory of hominids complicates the question, in that before grammatical language and math increased the options, the genetic or modular had been mixed with what is acquired piece by piece over time.

The encouraging thing about acquired habits and modularity in cooperation is that, if conventions can be disseminated and become more habitual, why not also more of the continuum and scalar spectrum? Because the measurements are extreme, they require attention and take effort to learn before we become familiar with them. But we can perhaps make them more habitual, more customary. That isn't easy; we can remember a week's worth of fictitious days of creation much easier than billions of years of accumulated atomic matter going through multiple stages. It is easier to have Adam and Eve rise up ready-made than to reconstruct a trail of cellular developments billions of years in the unfolding. Easier to load a homemade boat with a few token species than to account for millions of them, each with its own lengthy evolutionary history. The reason getting very far into unaccustomed distances, sizes, numbers, and temperatures receives so little help from modularity, to repeat, is that nothing in our ancestry prepared for temperatures in the billions of degrees and clouds of gas and dust trillions of miles across.

Magnification by itself becomes familiar with repeated use, but what appears in it may be as strange. A dust mite magnified a thousandfold, for instance, is odd looking even if we catch it having a drink of water like a household pet. We can adjust to minute and galactic things in books and on the internet where they come framed with explanations, but no amount of exposure to them entirely erases their peculiarity and in some cases their creepiness.

What we learn by rote is by definition not inborn, but inasmuch as one thing can be much like others, we are prepared to add a certain range of measurements to known ones. Likeness is the best aid for doing that, and it can also be useful for dimensions only vaguely familiar. That obviously grows less effective as the scalar extremes increase and counter intuitive anomalies creep in. For those, we fall back on foreshortened times and distances - on abstractions, models, samples, epitomes, mathematical compression, and quirky things like Escher drawings. We know what paradoxes, anomalies, conundrums, and perspectives-by-incongruity are, but applying them never comes as easily as applying commonplaces. Even a cursory glance at the CPEP chart of particles presented earlier tells us how much we have to learn the hard way and how much we miss because it is too small, too distant, too fleeting, or smothered in too much detail.

Pertinence

Together with discarding detail, singling out what counts accompanies comprehensive statements. I've been classifying this under the salience principle, usually connected to proximity in that the polar bear prowling a few yards off is momentarily more important to the seal than anything else. Every perception is self-oriented in that way, the invariable self being the formatted place-from-and-into-which. The variant self is you and I in our accumulated circumstances, one necessary aspect of which is placement.

Simulated and virtual placement are transportable and can be projected nearly anywhere. Samples close at hand can have the importance both of proximity and of what they reveal about comparable things. Using one moon for clues to the others raises its importance.

Even just one neutrino or one dark matter WIMP becomes valuable when it is measured and used as an example.

The importance of samples derives from that typicality and application. As we've seen, they gain in importance if they turn out to be epitomes. Within our customary range, something representative has automatic instructional value - take a chaparral hillside composed of rocks, brush, and gullies. Like any disrupted topography, it mixes the orderly and the disorderly as well as the organic and the inorganic. It gains in importance under the salience/proximity principle because no overview of nature is complete that excludes what it illustrates. Botanically, it offers a large sample of plants competing for space.

Geologically, it includes a rock history typical not only of the rock planet earth but of many others. Biologically, it includes predation in coyotes on the prowl and hawks circling overhead. Any selection of these factors yields general laws of the animal and vegetable kingdoms. The information this one contains in its minerals, flora, and fauna is voluminous. Its flowering plants alone fill a 300-page field guide, a characteristic sample of nature's high production. As usual, thermodynamic conduction, convection, and entropy apply to it as they do to all solids, viscous bodies, liquids, and gases. We can take that for granted in any sample. Time's irreversibility also applies as an invariable. None of the rocks that roll down the hill will ever roll back up again. None of the heat exchanges will ever decrease entropy.

Like much else, this landscape is subject to misinterpretation, but in western literature and philosophy places of greater height and extreme pitch tend to be more vulnerable to the invention of nonexistent forces. Take the Spirit Wordsworth places in "Simplon Pass." It wouldn't be assigned to anything as modest as a chaparral hillside:

> The immeasurable height
> Of woods decaying, never to be decayed, The stationary blasts of waterfalls,
> And in the narrow rent, at every turn,
> Winds thwarting winds bewildered and forlorn, The torrents shooting from the clear blue sky, The rocks that muttered close upon our ears, Black

> drizzling crags that spake by the wayside As if a voice were in them, the sick sight
> And giddy prospect of the raving stream,
> The unfettered clouds and region of the heavens, Tumult and peace, the darkness and the light–
> Were all like the workings of one mind, the features Of the same face, blossoms upon one tree, Characters of the great Apocalypse,
> The types and symbols of Eternity,
> Of first, and last, and midst, and without end. (3-20)

Despite that grandeur, the giddy prospect and raving stream aren't ideal illustrations of the mind Wordsworth imposes on them. Although saying "*as if* a voice were in them" backs away from an unqualified commitment to pantheism, it is nonetheless fanciful to think, even hypothetically, of black drizzling crags speaking. (Why *drizzling*, by the way?

Clouds drizzle but crags don't.) As we've been seeing, the tendency to animate the inanimate is typical of intellectual history nearly everywhere, in all eras and cultures.

Despite its representing much in little, an ordinary hillside discourages imposing types and symbols on it. It is true, however, that if one were to choose a time when light angles up from the west, the electromagnetic fireworks would move this one a notch closer to grandeur, actual and fully natural, not in any way animistic. So, let the reddish light of the declining sun dissect the slope, quilted by paths and fire roads. Let the electrified atmosphere spreading across neighboring hills, like energy made visible, go to work on its objects and contours. Before light reaches the dimming point, electromagnetism might then be imagined flashing across the surface and on out to sea. Wordsworth might find such a spectacle an intimation of immortality, but typicality is what gives it uplift. The "same laws of electrodynamics and optics will be valid for all frames of reference for which the equations of mechanics hold good," as Einstein puts one branch of relativity in "On the Electrodynamics of Moving Bodies," 952, 37). Or as Richard Wolfson (2003) phrases it, "The laws of physics are the same in all uniformly moving reference frames" (82), not grand in style but expansive and subject to verification. Placed in the continuum under the natural laws that govern them, a good many prospects which are not particularly auspicious in ordinary light take on added luster. Every object, however small, has that potential.

This particular scene gains that kind of status less in the way a fossil represents a species than in the way combinations speak to nature's mixed abundance. In James Lawrence Powell's words, "Threads from all fields [...] bind the tapestry [of Earth History] into a tight and interdependent cloth

[…] The tapestry of science crowns our species. To tamper with it is no different from tampering with another great tapestry of human invention: the arts. Imagine removing every fourth note from *Eine Kleine Nachmusik*, every fourth line from *Hamlet*, every fourth brush stroke from Van Gogh's *Irises*. To do so would be to deny and desecrate the finest that our species has achieved" (225). One critical difference is that where Mozart, Shakespeare, and Van Gogh put their compositions together intentionally and meaningfully, nature's combinations lack demonstrable management other than by universal laws. A good deal can be subtracted and added without radically altering nature's helter-skelter mixes. No craftsmanship carries from one moment to another as musical compositions do. Nonetheless, the links created by natural laws hold throughout. If you want to know how the chaparral hill got to be where and what it is, keep going back, not infinitely far but to the original singularity.

CHAPTER TEN

MOUNTAINS, RIVERS, FORESTS

The Argument

With their jagged masses, extreme weather, and great age, mountains make good scale extenders for those who place them in the context of planet history. They are assisted in that extension by their forests and streams. Not until the underlying tectonic plates were known could their depths be sounded. Post graduate McLaurin was acquiring enough earth science on his own to make them and what came with them just such scale extenders.

< >

When we try to pick out anything by itself, we find it hitched to everything else in the universe.

The great poets, philosophers, prophets, able men whose thought and deeds have moved the world, have come down from the mountains.

In such places standing alone on the mountaintop it is easy to realize that whatever special nests we make—leaves and moss like the marmots and birds, or tents or piled stone—we all dwell in a house of one room—the world with the firmament for its roof—and are sailing the celestial spaces without leaving any track. (John Muir)

< >

While the rivers of the ancient world brought the principle of circulation to settled societies–the Ascyron gifts of harvests, exchange, law, and empire–they were, at the same time, seen as the carriers of havoc and death. The Nilometers at Cairo were only auspicious if they registered exactly the right number of cubits. Too little, and one of Typhon's curses, drought, would visit the valley, bringing with it half the plagues of Old Egypt. (Simon Schama, *Landscape and Memory*, 260)

< >

Natural History Read Vertically

Topography illustrates the mixed order and chaos characteristic of atomic matter composites. No terrain does so more spectacularly than that of mountains. Take the chaparral hill of last chapter with its gullies, patchy vegetation, and animal life. Multiply its size and species tenfold to get a mountain, then remember that, even there, what has pushed it up isn't in view. Under the surface are uplifting forces reaching down much of the way to the planet's core. In vertical manifestations of the continuum, tectonic plates push down in subduction and volcanic plumes rise up. When geologists in the 18th century debated whether vulcanism or water had shaped earth's surface, both were partly right and both were unaware of the uplift and subsidence that Hutton soon proposed. Knowledge of nuclear and electromagnetic forces behind rock in a heated state was nearly two centuries away. Where it is a forest that stretches time, it isn't only in ancient trees but in species histories and in the soil as a combination repository and dispensary. Where a river does the scale extension, it is mainly by carving canyons and adding silt to valleys and deltas.

Before the nature writing of the 19th and 20th centuries, the openings that mountains wedged into extended natural history weren't nearly as detailed and demonstrative as they have become over the last century and a half. The atomism and chemistry of Sennert and Boyle in the 17th century looked into natural processes at the molecular level, but had no way to connect that level to species or to topography. Getting from the very small to the optical range and on to the gigantic wouldn't be possible for some time yet. Newton's universal laws didn't apply to the patchwork disorder of earth's crust or to how masses come together. In the next century, European and English neoclassicism avoided abruptness and extremes where it could and had little literary or philosophic interest in rugged terrain. The naturalists Hutton and Humboldt lacked knowledge of the atom and of what was combining and clashing deep within the planet. The fuller story couldn't be told until the advances in geophysics, astronomy, plate tectonics, and quantum mechanics of the 20th century. Any climb into cold, thin air, however, told mountaineers of natural selection's stringent requirements for survival. That much was always obvious. Living doesn't come easy in any marginal environment, but at steep, high elevations it comes the hardest outside of deserts and polar regions. Even at lower elevations, bears hibernate, insects hide in cracks and crevices, marmots dig in, and lean wolves, coyotes, and mountain lions prowl. On the upper slopes, where glacial ice permits, trees grow stunted and widely spaced. Torrents pour down through broken scree and around boulders.

Whitened tree trunks and branches stand as stark reminders of life spans. In the fallen 'mother' trees of lower slopes, life span limits and renewal come together in saplings growing out of prone trunks.

Except for the Eurasian expeditions of Marco Polo and John Mandeville, mountain ranges were known only locally until ships set sail. Not until the 17th century did western travel literature commonly include non-European terrain. Expeditions started mapping continents in the 15th century, but it took another two centuries to get shorelines placed on paper as they are in geography. Above the Indus, Tigris, Euphrates, and Nile, mountains and forests had been storing water and sending soil downstream for millions of years before agriculture got underway there. Released gradually from glaciers and snow fields, water was first captured for irrigation in those areas. Above the Willamette Valley and the Central Valley of California, the Cascades and Sierra distributed water to the first immigrants from Asia and, several thousand years later, to American and European migrants and settlers drawn westward by fertile soil, gold in the streams, and veins of gold, silver, and copper in the mountains. Before they arrived, life in the new world had gone unrecorded except for tales told orally.

Outside of the sublimity tradition stemming from Plotinus and the rhetoric instructor Longinus, relatively little that was favorable or informative about mountains appeared in classical, medieval, renaissance, or 18th century literature. As Marjorie Nicolson points out in *Mountain Gloom and Mountain Glory* (1959, 1997), the spotty literary history of the sublime had little science and not a great deal of descriptive prose. Classical and neoclassical commentary preferred symmetry and geometric proportions to irregularity.

Intelligence could be imagined behind well-shaped things, but only with difficulty behind vertical spires and deep canyons. In dark age allegories, enlightenment had usually meant indoctrination rather than assimilated and interpreted observation. Ascending a height signified spiritual growth through stages of initiation in a doctrinal context. Actual landscape features other than elevation weren't easily allegorized and couldn't yet be read in terms of mineralogy and distant stellar sources. As long as the elements remained just the four of Epicurus – earth, air, fire, and water – it wasn't possible to get very far into the dynamics of land mass and movement and the book of nature remained mostly unread (or misread). Even in Lucretius, despite the rugged Roman landscape and Mt Vesuvius not far off, none of the roughly 28 dozen times *De rerum natura* mentions fire and heat considers their function in opening volcanic vents and moving land masses.

Until the 20th century, travel and mapping were of little help in determining the stages of planet history's development. Nature was usually thought to have been created much as it is or as it soon became. That remained an assumption behind steady state earth history and astronomy well into the 20th century. Though mountain terrain could be hunted for game and used for defense, through most of written history it remained unexplored except locally and along trade routes. As Daston and Park (2001) point out, travelers and readers of their journals from 1150-1750 were more taken with marvels and freakish wonders than with the panoramic, the sublime, or of all things the mineral makeup of rock. The anthropomorphic habit was to look for purpose, even in irregular terrain. That the mountains had been there for millions of years before anyone saw them, and that what had pushed them up had been at work for billions more, wouldn't have been acceptable - even if someone pointed at sea and plant fossils at elevation and at the depths of eroded canyons and accumulation of soil in deltas. Grains of sand on the shorelines and in the seas told a similarly story.

A few generations before Wordsworth in the Third Earl of Shaftesbury (1671-1713) and eventually in the Romantic poets themselves, what had often been allegory in religious instruction or excrescence became magnificence if it rose high enough or plunged deeply enough. Wonders consisted more of mass and verticality than of oddities. English neoclassicism had recoiled less from the industrial revolution and commercialism than from what political moderates and conservatives thought of as the 'enthusiasm' behind the puritan turbulence of the preceding era. In the Romantics, mountains and rivers could be used to counter industry, urban life, manufacture, and trade. In their eyes, the grand and magnificent countered spirit-dampening materialism. The Alps and the English Lake District supplied all they needed for that. In England and Europe, knowledge of mountain ranges and rivers in Eurasia, Africa, and the Americas was scant.

In breaking with neoclassical constraints, the sciences associated with the Romantics, as illustrated by the Swede Jacob Berzelius (1770-1848) and the chemistry displays of Sir Humphry Davy (1778-1829), returned to the spectacular. Davy's demonstrations in electrolysis added to the marvels that in the poets, and later in Muir and Burroughs, came mostly in topography. For scientists, poets, and tourists alike, terrain still had some of the curiosities that had been there for Marco Polo and Mandeville, but they dropped the more bizarre combinations of vegetation and animal shapes and took note primarily of awesome expanses, elevations, and depths. This didn't rule out master spirits capable of shaping the terrain - the Transcendentalism of Emerson and Thoreau, for instance, had less to do with irregular terrain as habitat and vertical mass than with a presumed transcendent force behind it. Like Wordsworth's intangible power, that one also wasn't responsible for noxious and decaying things. The polar regions, jungles, and depths of the sea might as well not be there. Again, much like Wordsworth, Transcendentalism turned what under naturalism would be samples and epitomes into types and symbols. In the wake of that movement and the Romantics, the painter Albert Bierstadt (1830-1902) featured an outsized ruggedness that shrank foregrounded people and animals. Level areas led directly up against towering vertical rock faces.

With later contributions from Muir and others, the magnificence of scenic places in the Romantics, the Transcendentalists, and Bierstadt distinguished not an entire era but movements within the span of about a century that coincided with captains of industry and fortunes amassed in manufacture, trade, and transportation. The contrasts among movements were similar to the one between Davy's chemical wonders and the cloth-dying industry which came from them. The human impact on mountain ranges, not yet equipped with industrial level mechanization, was as slight as dynamited mine shafts and the naming of peaks. The latter was sometimes as unintentionally incongruous as George Vancouver's naming Mt Baker for a ship's officer and Mt Hood being named for a sailor in the Royal Navy. Such reductions of magnificence weren't unusual or recent. Proper names have frequently been incongruous, a nomenclature equivalent to Bierstadt's familiar plant and animal shapes looking minuscule beneath the rock wall heights of Yosemite.

The Spaniards had named the Cascades the "Snowy Mountains of St Antony," *Sierras Nevadas de s. Antonio,* despite no such saint having had anything to do with them. Hall J. Kelley would have had the entire Cascade Range become the President's Range on the grounds that it featured isolated eminences, thus not only mounts Washington, Adams, and Jefferson but Madison, Jackson, Tyler, Harrison, and Van Buren, the latter group about as fitting as Mt McKinley for Denali or, for that matter, the Three Sisters, Faith, Hope, and Charity with their attendant Bachelor and little Brother. As a further mark of the discrepancy between typical points of view and the wild, the durability of the name *Cascades* – so far extending to about two centuries – works out to only 2% of the time the range has had a title, and considerably less since earth movement started lifting the range at the north end some 36 million years ago.

Except for a handful of geographers, it took nearly a century and a half after Wordsworth's *The Prelude* for the vocabulary of plate movement – subduction, thrust and transverse faults, sunken graben, oceanic ridges, and the like – to translate into descriptive geology that which, outside of Hutton, had been picturesque or sublime. Common principles of breakage could be assumed, such as nature choosing the weakest place for breaks, and transform faults having two slabs slip past one another with less shaking of the earth than the collisions of thrust faults. Indeed, most naming and descriptions of mountain grandeur depended on keeping a proper distance and angle of vision to block off unwanted features, a characteristic device of anthropomorphism generally. Humboldt had accomplished a good deal before the mid-19th century but, outside of his and Hutton's mineralogy, common mountain vocabulary remained that of scenic grandeur. With a few exceptions, mountain, forest, and stream terminology still fell more under human downsizing than under natural philosophy and the physical, biological, and botanical sciences. Anyone searching for reality on the scale of mountains and river systems had to go outdoors, not into libraries.

At the subterranean rock level, the geophysicist Arthur Holmes, as we saw with respect to plate tectonics, wrote in the vocabulary of *eclogite* (red and green rock resulting from high-pressure metamorphism), *rheidity*, and *diapirs*, or intrusions of relatively light rock through overlying rocks while in the rheid or viscous state. The "buckling, bending and contortion of rocks" involve "fractures, sliding, shearing and flowage," in Holmes' terminology (133). Behind breakage at the weakest link is the overall thermonuclear principle of unequally heated areas generating movement and collisions. In contradiction to steady state theory, that principle preceded the forming of the planet by some nine billion years. Where the materials are solid and the movement powered by heat transfer, the mechanics bring uplift, sinking, and breakage. The formulas for rheidity can get complicated, but no other language for the shaping of mountains does as well for combined viscosity, pressure, and time. Nature's curved lines in ice, too, are owed to the ratio of fluidity to rigidity, under pressure, over time. River boat operators and plowing and discing farmers ply their trades atop something as foreign to them, and yet as elementary, as rheidity and rock-breaking diapirs.

Downslope Habitat

With their species proliferation and soil makeup plus tree branching, old growth forests illustrate multiplication by division as much as they do age and grandeur. In that respect, not much within our normal range models nature's productivity as well. The age and magnificence of trees are only modest scale extenders compared to mountains, but any excursion into an old growth forest shows how impressive they are on a human scale.

Their branching from mainline limbs into twigs makes a natural bifurcation chart. Some still living trees predate ancient civilizations. Chris Maser (1989) remarks that if calculations of its age are accurate, the General

Sherman sequoia "would have germinated in 1832 B.C. and . . . have been 632 years old when the Trojan War was fought in 1200

B.C. It would have been 1,617 years old when the Great Wall of China was built in 215 B.C." (10). For half the life of a mature Douglas fir, the Pacific Northwest lacked an extensively mapped shoreline. As products of sun, water, and soil, trees are chem labs. As repositories of wood-based products, they are warehouses. Forests are also the habitat of the predators that remind us again of Blake's "forests of the night." We need no longer ask, though, "what immortal hand or eye/ Dare frame thy fearful symmetry," since botany and biology determine growth, and "distant deeps or skies" take us into stellar dynamics.

Forest histories often contain incidents that connect civil affairs to natural history.

The forests of France and England harbored petty tyrants and rogues along with legendary green-clad heroes. Under Norman the Conqueror and his successors in England, forest administration favored noble venery that made wooded areas good places for hunters. In imagination, the native inhabitants of forests remained, to some extent, equivalent to children's scary stories of witches, trolls, and gremlins. By providing fuel, shelter, and hideouts from sheriffs, forests did often harbor a contrarian spirit if not actual strange creatures. For a couple of centuries, the North American forests enabled a future unlike that of older civilizations with their more limited wilderness patches. Tension between the landless and landowners was less critical where people could move into sparsely settled Territory, as pioneers did across North America for some two centuries. Lacking castles and defined domains, forests would have been roughly democratic even without providing places to hide from sheriffs.

A forest floor tells those who can decipher it a story based on chemical compounds and breakdown, plus the disintegration of rock on its prolonged course to grit.

As Roger Caras remarks in *The Forest* (1979), forest life works from soil up to the captured protein of predators: "The tangle of microscopic life [...] was a jungle unto itself [...] To creatures like the worm, lowly, yet large, all such life was a mash of organic matter to be ingested with the soil. Upon such material, the worm lived before it fed the mole that fed the shrew. In quantum leaps, the smallest motes of matter reached the largest and became part of it. This flow is called the tropic scale, and it is propelled by the power of the universe itself" (44). Therein lies a major value of playing scales: getting from one level to another, from the worm to the shrew and on. In the breakdown phase, the soil of a forest floor becomes a repository of decayed bits and pieces. Harvesting its information requires putting much of the succession together and understanding the relations to organic forms of non-sentient matter at the mineral and element level.

To those who come from crop lands and cities, a forest with a mountain setting has an uneasy side. Beginning with the bacterial mash, it harbors an assortment of devouring things, a ladder of appetites graded by size and prodded by chemistry, the force behind predation. As a place of devouring appetites, a forest is simultaneously productive and destructive. This isn't limited to predators devouring prey; scavenging, hunting, and gathering find much to collect and take in or carry home. What a scavenger comes across isn't intended for it, as when a plant that produces more seed than its replenishing needs keeps birds going. In terms of human consumption, relatively soon the biofuel that vegetation produces may replace a significant quantity of fossil fuel. If so, that will be a quick turnaround - first photosynthesis producing growth, then biofuel immediately converting it back into energy. Meanwhile, like a jungle, a forest is a thriving example of nature's productivity. Multiplication and division work into leaf and wood, in crawling, flying, and prowling life where competition and life cycles keep the populations down. In scavenging, the disintegration of something has only just begun before it is taken up and incorporated into another sequence. The use of unintended beneficial things like that is a major part of forest economy. The light and heat of solar nuclear fusion lands wherever it might and refuels the mechanism. As the planet's most ingenious and diverse scavengers, people harvest the fish and venison and convert trees into buildings.

Lifeforms overall, biological as well as botanical, tell a mixed tale of ecological cooperation and thorny defense. Bitter cherry thickets and gummy mountain misery don't lend themselves to romantic enthusiasm, nor does a tanglewood on the lower slopes of a mountain. The American Transcendentalists and European Romantics tended to ignore that, as ideologies and doctrines do as well. To those who have to force their way through it, however, a forest is randomness in growing, fallen, and decaying phases. Peattie reminds us that even colorful blossoms, which we think of as trigger mechanisms for hummingbirds and bees, aren't there for that pollinating purpose. They, too, are an excess. That is a major part of the story of mountains, forests, and streams in combination. Peattie puts that positively, but it has a wasteful side, on its largest celestial scale in the materials of nebulas that remain on the loose. His interest is mainly the flora and fauna: "Many flower-haunting insects seem blind to color. The beauty of a butterfly's wing, the beauty of all things, is not a slave to purpose, a drudge sold to futurity. It is excrescence, superabundance, random ebullience, and sheer delightful waste to be enjoyed in its own high right"

(*This Incomperable Lande*, 273). The more abundant the growth, the greater the disintegration - though the breakdown phase, too, sometimes has its use, as downed wood contributes to the forest floor and, in the meantime, provides nesting places for birds and hiding places for insects. Forest extravagance extends to such incidental things as golden mantle squirrels breaking speed limits before pulling up into statuesque poses. The squirrels may have some reason to move in that extravagant fashion, but avoidance of predators can't be it. Together with the slightly orange mantle, the flash of speed would attract attention and the pause would then make an elusive target into a sitting one. That's a diminutive example of what on a bigger scale is universal in "excrescence, superabundance, random ebullience."

That which nothing needs right away goes into a holding area (one way to think of the forest floor as well as nebular clouds and other kinds of limbo or holding areas). A breeze that lifts a fog or drives it through the trees like a fleeing ghost is a visible phase of excrescence. The presence of the fog helps some things and its disappearance others, but neither has anything to do with why fog comes and goes in the inter-working of temperature, moisture, and landscape. Similarly with nighthawks slashing and fluttering their skywriting at dusk, though the extravagant motion in that case, as with the quirky flight of swallows and bats, has a purpose in the pursuit of insects. Suspecting that something harmful may be nearby comes naturally to a forest dweller accustomed to undetected things appearing out of dim light. This is easier to imagine in dark woods than in most places although, despite stories of witches and ghouls, nothing unusual hides there, just an abundance of what is natural. Forests follow visible and well-known cycles of buildup and breakdown, played out in the turnover of flora and fauna. All the while, the mountains continue to make their age-old contribution by pushing up and breaking down into grit. Grit is a mountain in limbo, and beneath the chemical and molecular level the binding and breaking forces of the neutrons, protons, and electrons work away as always, everywhere.

Random ebullience applies to river courses as well, writing their chapters in plunging, foaming graphics, the spectacular part of water cycles. Pitch, obstruction, streambed composition, headwater glaciers, surrounding forests, and volume determine the pace and clarity of the water. Ripples, whitewater breaks, eddies, pools, surfaces, and stirred, bubbling depths create flow patterns that a hydrologist could chart only with difficulty, if at all. Mountain torrents throw themselves over and around boulders and off cliffs in ways that neither fish nor man can handle. A whitewater torrent is a moving image of excrescence and energy powered by gravity. Moss and roots take what they need of that plunging excess. We have no quantitative measurements for all its parts, only for volume, pace, temperature, and mineral content.

Where, literally speaking, streams are nothing more than enthusiastic water running downhill, Wordsworth makes them into a philosophical paradox in the "stationary blasts" of "Simplon Pass", constantly disintegrating while remaining in place. Flowing water also makes a good visual equivalent to the invisible flow of time and of things taking a one-way course to a prescribed end. Amplified to the maximum, that becomes the universe on its way to the still sea of entropy. It is after all true as Heraclitus says that we never set foot in the same river twice. As the water moves it wears down the rock bed.

Equally prominent is the intersection of differently paced courses. Last month's rainfall propelled downhill meets the boulder of 30 million years ago. This year's fawn carrying ancient quadruped genes treads on rock and soil of untold ages. Like everything else, that too grades down from cells, chemistry, and molecules to atomic minima.

What was traditionally portrayed in river literature granted less to ebullience than the real courses of rivers abundantly illustrate. Not all the inner workings of the great solvent water were known, and not much of the intersecting courses of rocks, creatures, and rivers - not even a great deal of the mechanics of gravity. In the17th century, under the spell of mysticism Henry Vaughan makes a waterfall into an emblem, simultaneously of the body's fall and soul's ascent. In translating 'death' into 'everlasting life,' he discards biology as well as nearly everything real in waterfalls. How a disembodied soul functions without sight, hearing, and touch isn't his concern, nor do obstructions and pools find a place in the metaphor. Instead, "Quickened by this deep and rocky grave," the soul merely rises "to a longer course." That's pretty much it. Where the course leads and what is there is left to the imagination, not only in Vaughan but in nearly everyone of his day and the preceding centuries. Another existence – anything but the real terrain of mountains, forests, and rivers – would be preferable apparently to what the senses report about the planet.

In a similar cleanup and discard operation, John Denham makes the Thames into a model, in his case not of souls going off but of good taste, thereby setting a neoclassical standard for writers minus the turbulence of whitewater rapids and of Puritan agitation. What he takes from the river is a model of restraint:

> O could I flow like thee, and make thy stream My great example, as it is my theme!
> Though deep, yet clear, though gentle, yet not dull, Strong without rage, without o'er flowing full.
> ("Cooper's Hill," 188-192)

Some such selection of what nature provides is as it must be, since poets who get too detailed and literal set themselves up for parody in the way of the Bulwer-Lytton yearly prize for mangled metaphor. In terms of Denham's preferred system of rule, being calm and gentle implies a monarchy presiding over a parliament satisfied to work within strictly assigned limits. Denham-style neo classical moderation sits at the opposite end of the poetic spectrum from Wordsworth's inserting a transcendent element into "Simplon Pass" - that too, of course, dispensing with most of what gives rivers their energy and force.

Let us return to that example for a moment. As I suggested, by using the word *like*, and thus only comparing the natural prospect to the workings of a great mind, Wordsworth avoids committing himself to a definable variant of anthropomorphism. Still, something spiritual has to be there if the rocky crags and the "raving stream" are to combine in an archetype. Consider the last portion of the passage again:

> the sick sight
> And giddy prospect of the raving stream,
> The unfettered clouds and region of the heavens, Tumult and peace, the darkness and the light–
> Were all like the workings of one mind, the features Of the same face, blossoms upon one tree, Characters of the great Apocalypse,
> The types and symbols of Eternity,
> Of first, and last, and midst, and without end. ("Simplon Pass" 3-20)

Converting natural heterogeneity into a single mind requires more than the usual detail suppression. "Characters" likens *things* to *words*, in this case forecasting an apocalyptic ending (one of the more prominent ways to end the world). The problem with the liturgical flourish is less that discarding of most of the world's inventory than the incompatibility, noted earlier, between the pass and the meaning Wordsworth assigns it. If it were actually the working of a mind, the giddy prospect, sick sight, and raving waters would show it to be unbalanced.

Declining to find either moderation or sublimity in a stream, Robert Frost is more modernist and modest in "West-Running Brook" in edging up to extended natural history:

> Our life runs down in sending up the clock. The brook runs down in sending up our life. The sun runs down in sending up the brook. And there is something sending up the sun.

Something confesses the limits of knowledge. That which has sent up the sun must also be running down, and this notion can now be carried further toward the singularity that sent up everything. Prior to the mid-19th century, accounts of mountains, forests, and rivers, and their emblematic uses, tell us how much a theory of evolution's long-term succession was needed to explain the flora and fauna part of the environment. As Richard Wilbur puts that segment of the continuum plus the "repugnant stone" of "Lamarck Elaborated,"

It was the song of doves begot the ear
And not the ear that first conceived of sound:
That organ bloomed in vibrant atmosphere, As music conjured Ilium from the ground.

The yielding water, the repugnant stone, The poisoned berry and the flaring rose Attired in sense the tactless finger-bone
And set the taste-buds and inspired the nose.

Inspired is an inspired word choice, and *yielding, vibrant,* and *repugnant* not far behind. Similar biological extensions of the continuum drive the legs of centipedes and tune the ears of owls to rustling in the dark. The wordplay and imagery recall former traditions; the myth of Amphion, for instance, and the music of creation in Marvell, Milton, and Dryden (Marvell in "Music's Empire," Milton in the creation books of *Paradise Lost,* Dryden in "A Song for St. Cecilia's Day").

Prospects

It isn't absolutely necessary to follow our model case history through from *H* to the extended vistas of *V* by way of the scale extensions mountainous terrain encourages, but, as these examples illustrate, the literature of rivers and mountains takes detours away from what is accountable to the landscapes. Atomic matter history hasn't been as profitably looked into, even now, as it could be. So, let us briefly turn back to McLaurin for that limited purpose. He also has to work free of common narratives to get at something more demonstrable. His first memorable close-up exposure to the vertical, steep, high rock of mountains came in the company of a fellow youth as impervious as he to anything not at hand and not subject to myths. Testing the west side of a mountain in the Oregon Cascades, they had reached about the 7000-foot level, with 3000 left to go, when they took opposite ways around a trapezoidal island of stunted trees. Finding himself alone at the top of it, McLaurin labored up a patch of snow-free shale toward a ridge that looked negotiable to the peak's basaltic spire. That final bit would have to be left for another day, after technical instruction and with better equipment, but he saw nothing to prevent his reaching the base of it.

Being out of contact with anything familiar, even at that modest elevation, was putting him in touch with something he couldn't name but felt deep down; it made him anxious. Although he later came to see more in that precipitous situation than he did at the time, even then it was clear that such places weren't for humans. Realization of that came on strong when he looked west and saw dark purplish clouds tumbling in, one over another, and realized that a storm was about to add one too many complications. If he didn't slide over the bone-crunching drop where the shale ended, he would likely become lost in a blizzard. A two-footed creature of compacted weight didn't belong in the habitat of eagles. Once that had his full attention, he retreated (as his partner, apparently, already had). Encountering a rogue wave at sea might have served almost as well for a lesson in natural hazards.

On and alongside rivers, he felt less uneasy - but plunging falls and torrents, too, were best seen from a safe perch. As he eventually found, rivers had attracted a commentary more voluminous and favorable than that of mountains. In its mythic chapters, it included strange beings, from satyrs and nymphs to Oceanus and his wife Tethys, mother of rivers. Fresh water, navigable courses, and irrigation were among the reasons. In their lower regions, rivers were to untamed nature about what domesticated animals were to their wild ancestry. He came to know some of the literature in due course, but to get a personal sense of rivers required going up and down them from valleys and outlets to glacial origins. Forays into extended scales demanded both footpaths and libraries. As a student of topography exploring the drainage of an area – using Google Earth where he hadn't personally trekked, swam, or rafted – he eventually followed the north San Joaquin down from the high Sierra. It initially pooled in Thousand Island Lake beneath Banner and Ritter, and from there worked its way down to Central Valley, an agricultural treasure 450 miles long. He picked that particular river because of its prominence in Muir's Sierra and its similarity to other westside drainage systems from Alaska to Panama and on down from there along the Andes - in other words, almost pole to pole. The upper portions descended from the vertical rock face on which Muir found himself stuck, unable to go up or down.

The upper river had all the unnavigable turbulence, riffling, and deep pooling a student of mountain rivers could want. From snow melt, it gained enough volume to join the Sacramento in flushing out the tidal basin and San Francisco Bay. These two rivers combined were second only to the Columbia among the westside drainage systems that he came to know personally.

The San Joaquin by itself, without the Sacramento, put in abbreviated form what were thousand-mile channels elsewhere, the world's largest in volume being the Amazon with many times the San Joaquin's vegetation, animals, and volume. In former days, the San Joaquin had provided a fishery and waterway for inland tribes; it now supported irrigation and shipping as far inland as Stockton. Everything the upper river did depended on upper-level rock, snow fields, and glacial water released in small amounts at a time. In most of these features, it was typical of drainage systems around the world. Similar systems had supported foragers and farmers on the Eurasian continent: in Mesopotamia, Egypt, South America, and indeed wherever humans wandered in their original migrations around and out of Africa. Some channels in the West which returned water to the sea were so deep that they were visible from outer space. Two of them, the Snake River Canyon and the Grand Canyon, both familiar to him in due course, were among the deepest in the world. The early migrations had followed where fresh water led, testing climate, flora, and fauna.

Long before record keeping got underway, the valleys of the Tigris, Euphrates, Nile, and Indus sponsored the surplus needed to build civilizations. They provided fish, floated tonnage, and removed waste. Their deltas had long been depositories of upstream soil that had used the water course for transport. The lowlands and valleys of China and Southeast Asia had as well. of course, but he knew little about them.

Such fresh water courses were coolants and habitats as well as drainage. As arterial conveyances, they served plains, deserts, and corridors of flora and fauna. No wonder mythologies chose them for the ever-elusive fountain of youth. Even after exploration had mapped them and transportation started making use of them, stories of marvels still clung to them. As the run-and-fetch-it servants of their circulation, air currents were integral parts of them. That all aspects of the planet were intertwined was nowhere clearer than in clouds, water movement, and sea currents captured in satellite photography. Because of the water circulation and air flow through passes and over mountaintops, nothing could be set apart decisively. Directly above rainforests, the air extracted water, lifted it, and immediately dropped it in the most locally contained of the water cycles. The energy and enabling powers were no longer secrets; everything in sight put electromagnetic and gravitational forces on display.

In addition to floods, river history came with drought, pestilence, and war. Typhon claimed certain prerogatives alongside Osiris in what Simon Schama (1995), with a flourish, called the fluvial myths of hydrological cycles. None of the rivers of western North America had comparable recorded histories, but they had been no less important to indigenous people over about the same dozen millennia that saw the rise of planting and harvesting, trade, harbors, monumental architecture, and irrigation. Rivers ran through the histories of all early civilizations. Thanks to the wealth they enabled, they encouraged invasions that tied them to eventful history. Wherever people recorded what they had done, rivers ran through the accounts, first in the 'between the rivers' countries of the Mesopotamian delta and along the Nile, and then, apparently independently, in Asia and eventually in oral traditions the Americas.

As important as the rivers or more so, mountains also impressed those who were getting initiated into what topography revealed of deep history. To hook up with them was to send the mind scurrying. McLaurin understood Cortes' men in Keats looking from a mountainside, "silent upon a peak in Darien," in the first European sighting of the Pacific, at the end of their land progress and beginning of a new phase of oceanic thought. If he was to grasp more of the whole – underwater, aloft, in flat and in rough terrain – it had to somehow enter the blood stream directly, as intuitively as understanding the weight of a stone by lifting it. Animals of all kinds took shelter from wildness and tested it cautiously, like marmots sniffing from the safety of their boroughs. He knew something about that, too, drawing in and peering cautiously out - cautiously to be sure but, as he reminded himself, as a creature of mountains and their immensity.

CHAPTER ELEVEN

HOME GROUND

The Argument

Through electronic and printed means, typical households now have access to unprecedented quantities of information and theory, and to simulated voyages in among molecules and out into space. One of McLaurin's former home areas had discards strewn about, ranging from outdated machinery to stone age artifacts. Put together with terrain and species, they made a track that could be followed into a reasonably deep past.

< >

It was here. This was the setting and the time Of year. Here in his house and in his room,
In his chair, the most tranquil thought grew peaked. . . .
Wallace Stevens, "A Quiet Normal Life"

< >

Home Delivery

Now as opposite to mountainous terrain and footpath exploration as common experience allows: ranging abroad while staying at home. Thanks to AI and television, simulated travel can go outward and backward as well as down in among the quarks, protons, and molecules. The simulations and samples are mind boggling in extent and variety. Fortunately, they all lead back to an initial singularity and have followed unswerving laws. Second only to that increase in the scope and convenience of what can be summoned is the ease of cross referencing and relating one thing to another. This can now be done with some confidence if not with absolute certainty under all conditions.

That not many householders take advantage of these excursions and adjustments in perspective is largely due, I think, to the attention and curiosity of most people being otherwise diverted to more pressing affairs.

Familiarity with extended natural history isn't a necessity for most purposes, although it is, in my opinion, for collective long-term survival on a besieged planet and for checking the reliability of long-term schemes and beliefs.

Home ground is a flexible term. For individuals and families, it retains the importance it had for Cro-Magnon man standing at the mouth of a cave. For citizens, it usually means a native or adopted country. For someone aware of the continuum, it can mean planet earth. If a multiverse exists it might even mean the visible universe, though outside of science fiction that would obviously be a considerable stretch. For nearly everyone who isn't homeless, it means nothing more involved than one's habitual dwelling. What I mean by it in this context is the place into which data and ideas arrive, a learning center focusing on the continuum. For the curious, almost anything recorded in any period or simulated at any phase of the universe can be summoned to this house, this room. Both print and electronic transmission put incoming materials into compressed forms. Two chapters ago, I considered the means of bundling to be what gets sent forth. Here, the concern isn't the dispatching but the receiving.

The potential clutter of that increased volume requires a good deal of sifting and sorting. Points of view on the evening news arriving by radio, television, cell phones, and computers skip in a few seconds from typhoon wreckage in the Phillippines to the floor of Wall Street. The air waves are also buzzing with misconceptions and fantastic alternative universes. Lack of focus, plus the usual deluge of the contrived and the manipulated, is the negative side of that expanded ready access. A tweet or blog sent forth at noon can become a common opinion by evening. Mixed in with the dramatized, the virtual, and the fake, the real does have to be sorted out more carefully now than ever.

Summoning so much so easily isn't the solely product of technology; the economic and social revolution which got underway in the 17th century contributed to it, in that it brought rapid population expansion and the larger and more varied consumer class that the industrial revolution needed. The delivery system wasn't limited to sources of information but included all manner of household goods and services. At both the individual and national level, it also includes weapons. Over time, learning has become democratized and now includes a barrage of distractions. To study under notable scholars, students in Donne's time – when making one room an everywhere was hyperbole for the presence of a beloved – had to travel to sources of instruction. Thanks to the devices that came with print and harnessed electricity, a typical research trail now goes from an index or a search entry to an information collection center. Locating a specific matter takes an elapsed time of a few seconds. While the total distance traveled physically is negligible, the total distance covered electronically is variable up to billions of light years. Most likely, the way to a source will be littered with attention demanding distractions, comparable to printed ads stuck over newspaper headlines and to unwanted, monitor-crowding inserts posted alongside relevant search engine results.

In the context of playing scales, the question is what this increased accessibility means for exposure to the continuum and the scalar spectrum. Almost anyone who isn't homeless and wants access can manage it, provided only that transmitters and recipients share a common language. Over the past half century, the number of information sources in total has raised a fear that snippets, bits from here and there, will displace the fuller working out of topics. Though the contrast between casual gossip and instruction is nothing new, the trivia of some 200 chatty websites with logos and subscribers keeps that fear alive, for it is capable of smothering what counts in natural history. Meanwhile, however, the usual histories, news reports, and books do still continue. Reasonably realistic accounts of periods and places are more numerous than ever, in literary historical fiction, in documentaries, and in filmed reconstructions - that goes for deep history visible in the celestial scans as well as for conventional historical eras like the French Revolution.

Extravagant Tourist Venues

Extra vagant: *wandering wide*. Traveling abroad, as I suggested, can now be done without leaving home, mainly in books and electronic virtual travel. It always could be achieved through imagination and reading, but is now assisted visual aids that come with commentary. Antarctic's Ross Island and the company of Erebus, offspring of Chaos, seem almost homey compared to other planets and alien galaxies on display in virtual travel, which has the added advantage of a guaranteed safe return. If a manned spacecraft ever does leave the planet behind for good, losing contact with the home place will be one concern. The perils out there are another. Life on one of Saturn's moons would be not a little intimidating. In *Death from the Skies!* Philip Plait (2009) reminds us of what lies in store for actual space travel: "The Universe is an incredibly hostile place for life. Virtually all of it is a vacuum, so that's bad right from the start. Of the extremely few places that aren't hard vacuum, most are too hot for chemical reactions to do very well–molecules get blasted apart before they can even properly form. Of the places that aren't too hot, most are too cold–reactions happen too slowly to get interesting things to occur" (from the Introduction). Movie shots of muskets poking out from overturned Ticonderoga wagons and arrows stuck into the ground seem familiar and homey compared to places off the planet. Even an expedition from a European nation to the natives of Borneo is moderate next to visiting Jupiter. All such voyaging and exploration can now be reduced to simulated form and taken in at home without the expense and the peril.

The comfort of knowing that the same basic laws of physics and chemistry apply everywhere is limited where water and air are missing and temperatures range from no heat to millions of degrees. At home, we can remain at ease with flyby views of Jupiter, an object of beauty from a distance that makes a storm on its surface the size of earth look to be no more than a reddish spot. Venus, too, looks bright and beautiful from the distance of a camera sent abroad, its images posted on televised science programs. Its greenhouse gas-trapped surface heat at 860°F would suffocate anything within reach that has to crawl, walk, or fly. Space travel continues to come in print as well as on monitors, in Barrow's *Cosmic Imagery* (2008), for instance, in Pauwels' *Visual Cultures of Science* (2006), and in Daston and Park (1998). The clarity of telescopic images improved significantly with the placing of telescopes and cameras in orbit and sent on missions around the solar system. It will again when the James Webb goes up, at the moment projected for 2021. Fields of radio wave disks have made similar improvements using the means of detection that scans the CMB. We've no need to go that far to find truly weird things, however - not when something of a twilight zone dwells inside us. Brendan Buhler (2008) remarks in "The Teeming Metropolis of You" that 90% of us is composed of microbes. In the small intestine, the human body "is a complex ambulatory feeding mechanism for a methane reactor" (3). If we could shrink enough to enter a living organism, its insides would seem not a little scary. Landing on Mars might seem preferable.

On the micro scale, in the canyons of the mind as well as in the body's teeming metropolis, we carry strangeness around with us. Everyone is a specimen filled with more pieces than the cosmos has stars, which we're told are in number comparable to grains of sand on the planet. If we are truly at home with anything, it should be our own brains. When seen from another angle, however, they also become foreign, all 80 billion neurons of them threaded through a maze of pathways. Getting immersed in any extreme increases the desire to get back to normal where gorillas are large and ants small. We choose to see what we would normally avoid, only if it is framed for convenient viewing, as it can be in living room expeditions. There, it comes with calm voice-over commentary that converts simulated movement and strange sights into connected information. Voyaging and penetrating thought work much the same in everyone; poets, philosophers, scientists, technicians, and school children. Almost anything anywhere can be delivered in such an informational context suitable for home delivery. We sign for it, unwrap it, and examine it.

Clutter

Anything not put in a meaningful context and assimilated seems intrusive and extraneous to the intellect, as in experiments anything not bearing on a theory is noise. Noise in the sound spectrum is equivalent to turbulence in other forms: twisted movement, blinding and wavering light, kinked and knotted thread, random temperature fluctuations. These are troubling to the intellect and generally unavailable to science.

This doesn't necessarily mean that the noise is from something actually random, but it might as well be from the standpoint of how it is perceived. Pertinence is measured by what fits a given framework and is under control. We can do much of the filtering effectively ourselves with the control that comes with common office and household equipment.

Encountered in the wild, nature contains a good deal that needs sorting out before we can do much with it. That's a universal phenomenon. Some dislocation and semi-controlled chaos, as K. Tomita (Holden, 1986) regards them, underlie "almost all natural phenomena." What doesn't hang together is neither "exceptional nor peripheral" (211).

O. E. Rustler (Holden, 1986) in citing Anaxagoras (who in Fragment 12 invented a non- existent *nous* to have a place free of what troubles earth) likens the control we impose on nature to putting containment around colliding plastic balls viewed from the outside. When we do have to plow through confusion, we prefer to take a navigated course with a destination and to get there as quickly as possible. The chief advantage of simulated voyaging I mentioned earlier: the ease of closing down, getting out of Borneo or away from Pluto and returning to the chair in the room, in possession of information we didn't have before.

The amount of detail in even a limited area would be overwhelming if we couldn't filter out the noise. The decision to be receptive doesn't necessarily depend on how pleasant or coherent something inherently is, however. In terms of art, modernism went in for controlled clutter in Pointillist and Divisionist technique and the prose of James Joyce's *Finnegans Wake,* the source of the word *quark*. Whirlwind images can be set to music patterned by rhythm and melody and thus made admissible, to an acquired taste even pleasant, as Mark Heard's "Another Day in Limbo" presents Indian and American ancient and modern desert scenes in a place that, despite its angels with dirty faces, is neither heaven nor hell. A list like Richard Shindell's crowd of vehicles leaving a metropolis ("Transit") is an urban equivalent likewise set to music, a miscellany contrasting to a choir practicing in a prison where assorted felons momentarily become a harmonious group. Clutter under containment placed across from something orderly becomes part of a set, each side of which keeps to its place. This is much like the wild across a fence from a cultivated field, one of the basics McLaurin acquires early on. Multi use compounds (hospitals, shopping centers, and business towers, for instance) have some of each, the confusion and the containment. A hundred people on the streets of Paris can be crisscrossing on errands in a general confusion, but each course by itself is purposeful and patterned, as is the grid of streets. Every vehicle keeps to its side of the street, thus order amid confusion or confusion locked into a grid.

Except in simulated and limited form, miscellanies aren't usually allowed in places under control. In its escape from nearly every means of location, limbo is the exact opposite of home ground. It has no predominant direction or means of relating one thing to another, no intent or goal, no patterned velocity or vector, no geometry, no landing places where flying objects can come to a stop. In short, it lacks a systematic grid and measurements of distances and angles. Even velocity means very little when the movement has no definable direction or duration. Name any kind of order except momentum on a course, and limbo will be found not to have it. Except for energy and movement, it is a comprehensive negative, further from home than anyone wants to be.

Home ground is so completely opposite that it might seem surprising to find representations of limbo being admitted there. The key to its being accepted lies in the art of its representation and its containment. We have no trouble listening to "Another Day in Limbo," "Transit," or an equivalent because the performance is of a *concept* of limbo in a rhythmic and tuneful form, not an imitation of what it actually feels like to be trapped in a melee. The most scrambled painting ever made is contained within a frame. The more comfortable packaging of foreign places is by means of language and foreshortening or zooming. Generalities also smooth out surfaces and establish orderly relations among categories. The purpose they serve with respect to data is similar to that of a grid with respect to terrain; using generalities we can make orderly, meaningful comments about limbo and chaos, as I hope I've just done. Assigning a name to the visible universe such as 'the cosmos' is the ultimate click of the mechanism, a semantic still shot that arrests the motion. Supreme fictions manage that with nonexistent places. I glanced earlier at a version of one made a final destination for selected people. Most of natural history and all of limbo aren't allowed in it. Such places aren't permitted to have jagged peaks, ravines, intrusive viruses, and beasts of prey. 'Going home' to such places in gospel songs turns planet earth itself into a prolonged dislocation.

Since people aren't very good at establishing and maintaining permanent sanctuaries themselves, supreme fictions are usually turned over to supernatural powers. Lesser cleanup we assign to abstractions, since, as Stevens' "Connoisseur of Chaos" puts it, in close "the squirming facts" would otherwise "exceed the squamous mind" (207). In "Notes Toward a Supreme Fiction" that is one reason for the initial premise, "It Must Be Abstract." The connoisseur of chaos, incidentally, was there about three decades (1942) before chaos theory got underway in academia. By the connoisseur's time, narratives, poetry, and dramatic versions of the scattered and the confused were popular as Thomas Beckett, James Joyce, T. S. Eliot, Ezra Pound, and other modernists thought of them.

Quite a few educated households acquired something of that disarray in print, paintings, and atonal music, which implied something along the lines of the Connoisseur's contradictory opening declaration:

> A. A violent order is disorder; and
> B. A great disorder is an order. These two things are one. (Pages of illustrations.)

Changing the perspective could conceivably make the nonsense of 'these two things are one' a meaningful proposition. In any case, the pensive man as the source of such a proposition, whether at home or at large, "sees that eagle float/ For which the intricate Alps are a single nest." Distance and soaring altitude are what make multiple spires of rock look like a nest.

As we saw last chapter, and in several places earlier, species names are similar in clearing away details and citing collections of like things without concern for variants. To return to a prominent example valid across the universe: calculations of the multi-directional gravitational pull on any given body have to be simplified, otherwise too many masses tugging on too many others would have to be computed. Even with just three, locating the gravitational center isn't easy, but at least the distant weak tug of Jupiter can be ignored on Earth, as can many another too faint to matter. In the language of gravitational studies, "the general motion of many (*n*) bodies interacting through gravity . . . is not integrable" (Carl Murray, 1993, in Nina Hall, 98). Sighting objects seen as wholes presupposes distance from them. In amongst their molecules, they are raucous - much as amongst the intricate Alps the terrain is steep and confusing.

Getting a comfortable distance and thus a perspective on something isn't the same as being at home with it, but it is related in the search for a perspective that works.

Whatever songs, poems, religions, and philosophies say, no one can be at home in an open universe. Too much of it exists and too much is in confusion to assign it a design. To be at home with something, we need it to have design and a location. A spaceship traveling at a high speed moves in point of view so rapidly that such a targeted object, used for orientation, has to be distant. The only objects actually at home to passengers in one are what travels with them. Before liftoff, targeting a landing place for a temporary home requires multiple calculations. Earth is moving, the projected landing place is moving, the space craft will be moving, and gravity will be working at different strengths from different directions on that ever-changing course. Within the common sensory range, looking out of a speeding car tests that principle, the car becoming a temporary home in motion (or, more precisely, in multiple motions) with respect to other objects on the planet, on the planet with respect to the sun and moon, and in the movement of the Milky Way with respect to Andromeda.

Home, Work, and Institutions Historically

A habitual location can provide not only the seclusion needed for concentration, but continuity up to a point. Regarding the acclimated "unity of consciousness" behind that, Nicolas Humphrey (2008) finds that his coming together upon awakening is assisted by it happening in a familiar place. That seems to be typical. Comfort and the elusive thing we call happiness depend to a large extent on such a location, missing in the homeless, refugees, and troops at war. Outside of aristocratic circles, private living quarters were out of the question for most people under monarchies and oligarchies. As Joseph Amato points out in *Dust*, medieval peasant hovels allowed the invasion of nearly everything except formal learning. The presiding cultural habits and institutions deprived all but an elite few of literacy. Medieval peasants "had no gauges to record pressure, no meters to measure power, no detectors for smoke or gases, and no thermometers" (38). Also, no books or the capacity to read them and not much control over their homelands. Under those circumstances, the ability to place things logically and quantitatively was obviously quite limited. Hence peasants, Amato concludes, had no means of calculating "volume or speed or ascertaining conversions between liquids, solids, and gases. Belonging to a rude and coarse world, overshadowed by dust and darkness, they had neither desire nor knowledge to manipulate minute things" (38). What a hovel dweller thought he glimpsed of an alternative place or a comprehensive order beyond king and country, others had planted in his mind at a young age.

That would have been true of most, if not all, cultures and political systems. Contact with that alternative place would ordinarily be only through one of the suborders of the kingdom itself, thus escaped only by dying.

In the West, the lack of a home place was historically the lower class condition until the social upheaval that accompanied the industrial revolution and carried empiricism along with it. Or perhaps better said in reverse: the upheaval that accompanied empiricism carried the industrial revolution, democracy, and education along with it.

Education was never a problem for the few for whom investigation came primarily in living quarters and remained independent where it didn't require patronage. That had been the case for only a short time before Bacon's utopian House of Salomon in *New Atlantis* (1627), where areas of investigation occupy separate chambers of health, weather, mechanical arts, sound, smells, math, geometry, deceit of the senses, motion, and more. Though called a house, it is more like a research complex that today would occupy a cluster of buildings on a campus, separate but pursuing a common educational venture with limited interdisciplinary connections. What passed for orientation differed by denominations and nationalities but was, broadly speaking, based on a corruptible earth, categorically and conceptually distinct from uncorrupted heavens. The latter were organized in rotating spheres made of an ethereal, or as Wikipedia says, of a transparent "fifth element (quintessence), like jewels set in orbs." In other words, nature had nothing among its known elements capable of such order and permanence.

Well into the 17th century, telescopes and microscopes were built as well as used by individuals working more or less on their own either at home or close to it. No factory polished the lens that Hans Lippershey (1608) installed in the magnifying instrument which preceded Galileo's telescope. Beginning as early as the 15th century in Poland, some learned societies could be called in-between forms for the furtherance of science and philosophy, as the Socratic dialogue groups of Greece had been as well. They weren't isolated or, strictly speaking, domestic, but neither were they funded institutions equivalent to public and private universities. The Puritans maintained that reading scripture at home or in chapels needed no guidance from bishops, high church ceremony and sacraments, or learned commentary. The Holy Spirit came directly in, usually received and interpreted by the head of the household.

Some of the studies that individual initiative undertook in the later 17th century led to university appointments. Newton's career paralleled the development of physics at Cambridge, though theories from various sources remained private enough that it was sometimes difficult to say who came upon an idea first. That lack of an international data exchange contributed to the friction between Newton and Leibniz over the invention of calculus and between him and Hooke over whether light was composed of particles (Newton) or waves (Hooke). The study of Latin and Greek had been a curricular mainstay for some time, but the compiling of English dictionaries was no exception to the new wave of homework. The first to put English words in alphabetical order with definitions was Robert Cawdrey in 1604. A century and a half later (1755), Samuel Johnson's dictionary, too, though handsomely sponsored by London booksellers, was compiled with minimal clerical help and not in a university or publishing house setting.

The encyclopedists of the 18th century remained largely independent, as did the Jesuit Roger Boscovich (1763) and the multi-faceted Prussian naturalist Humboldt in the early to mid-19th century. A philological society, another in-between organization, got underway at the University of London in 1830 to foster the study of language structure and history. The Oxford Dictionary was a mid-19th century academic undertaking. In the early 20th century, expeditions like those of John Franklin and George W. DeLong to the north and Shackleton, Amundsen, and Scott to the south remained privately organized and funded despite being matters of national pride and having naval connections.

As studies in natural history moved into an age of precision manufacture, settings for observing and testing like the observatory on Mt Wilson, home of the first generally accepted demonstration that galaxies exist, were assigned to research centers that needed substantial funding. Participants worked not in living quarters but in organized, equipped research areas. Mid to later 20th century organizations like NASA, Fermilab, CERN, Argonne, Oak Ridge, and the Gran Sasso National Laboratory in Italy were the beneficiaries of sizeable grants and government allocations. It isn't possible to catch a neutrino without a substantial project and an organized research unit with a suitable administrative structure. To put a telescope into orbit requires considerable NASA enterprise. Modern telescopes are complex, expensive, and increasingly clear in detail as are microscopes and electronic means of detection. Those that are set to catch neutrinos streaming from the sun are very elaborate and have to be buried deep enough to filter out other things flying through space. At Fermilab outside of Chicago, particles that have broken loose are caught on the move by technicians and scientists organized as teams equipped with detectors. In a tunnel to one side of the Gran Sasso lab, a XENON10 detector staffed by thirty scientists was set up to catch the elusive dark matter WIMP in the act of knocking into a Xenon atom. The staff and visitors at CERN are numerous and the organization that of a city unto itself, with sections for finance, computers, international relations, and many others. Team members carry on as an organized workforce using means of representation and transmission that send the results to multiple places, another node-with-strings structure of information collecting and dissemination.

As such multinational institutions show, it is the research this collects, the staffs, and

the continuum overall that directs the research. By printed and electronic means, the findings can then be sent wherever summoned; to television sets, cell phones, books, and PCs. Classrooms and home learning alike benefit from streaming and DVDs. Teachers from scattered places get invited to lecture recording centers. As a consequence, the telecasting of the outer solar system travels of New Horizons, especially its Ultima Thule flyby, became almost as noteworthy as the discovery of the Higgs boson (2012) and just as available for home delivery. (*Ultima Thule* is the name given a small object in the outer solar system shaped like two lumpy spheres crammed together, important as a sample of the newly forming solar system.) Before such large organizations as NASA, CERN, and Fermilab were in place and before the common transmission of their findings, the usual way for like-minded researchers to assemble was at conferences.

These still meet in specialized areas of research, none more impressively than the Solvay conferences that convened the leading physicists of the earlier 20th century. There, in a single building (temporarily their headquarters), they mingled casually as well as meeting purposefully.

As an overall international enterprise, modern research now collects into filamentary networks bunched into convergence points. Under improved methods and equipment, and in the transmission of data and theory, that exchange differs considerably from any preceding research enterprise. Such cluster-and-string configurations come naturally to multi-unit coordination lacking a command center. Telescopic views and photographic recordings of galaxy clusters find similar networks stringing into and out of denser clusters. Cognition studies find that the brain, too, has neurological networks and meeting points rather than a dominant center. This doesn't into social or political clusters carry in any straightforward way, but it does resemble republics and democracies more than authoritarian systems. Equally dispersed and connected are multi-campus university systems like those of New York, Texas, and California. They have central administrations but turn curriculum and research programs over to local management, subject to system wide review. Many of the 'strings' go from campus to campus.

The concept of home ground and nodal clusters carries well beyond human reception centers; it comes built into natural history's own density centers with their tenuous connections. Gravity is assigned a core in each mass, and several in combination resemble string and cluster configurations. As a parallel to a home center drawing in information, this raises something of a conundrum in positioning segments in two places at once – that is like the surface of earth being 4000 miles from the core and yet contributing to the planet's gravitational attraction as if *from* there, or from wherever the core has shifted in response to other attractions. It simultaneously pulls and is pulled.

One wrinkle in that oddity is the relocating of the core itself according to the strength of an opposing attraction. The planet's gravitational center is drawn less toward the moon's than the moon's is toward the planet. The gravitational center of two masses can be computed, and even those of three, but not of a dozen of different sizes and distances, all of them moving. That involves other matters as well that we needn't go into, such as a segment's location in each mass being attracted at a strength proportionate to its distance from the core, full strength on the perimeter, less and less going inward, all that too shifting in response to other gravitational masses. The likeness to a home ground learning center is faint but, I think, genuine: situated in such a center, a student of reality is drawn to dispersed centers of learning and information. These, in turn, attract and disperse data from all over. The clusters have convergence points, such as someone's home address, and the connectors streaming into and out have ideas and information passing through.

Refugees and Pilgrims

Establishing or enlarging a territory for one group usually means losing it for another. This can turn entire populations into refugees looking for somewhere to land. I commented on that earlier, but it warrants further attention in the context of home ground and its webwork connections. Refugee victims of imperialism lacking both of these may carry convictions, learning, personal relations, and loyalties along with them, but they lack settled geographic placement. Once migrating clans out of Africa had reached nearly every habitable place on the planet except the Americas, skirmishes over territory probably began - so the estimated age of some archaic weapons suggests. (Before bronze weapons, copper was in demand for maces, probably not used for hunting game or not exclusively for that.) Those in possession of a territory had to be moved off before incoming migrants could take over. A century after the bronze revolution of the proto dynastic period (about 3150), with bronze adding strength to copper and tin, the troops of Sargon of Akkadia (2334-2279 BCE), 'true king' or 'Sargon the Great' as flattering historians called him, used improved weapons to consolidate Mesopotamian cities into an empire (a nation consolidates city states. An empire consolidates nations.)

For several centuries, the predominant form of contention over land featured European immigrants and imperialist powers moving into selected areas of Oceania, Africa, and, beginning in the 16th century, the Americas. Resources and markets were the economic driving forces, but imperialist powers did so also in the name of progress and enlightenment. Whatever the particular motives of a given expedition, nearly every European country with seaports sent expeditions abroad. If populations of old resembled those we know from the past few centuries under improved record keeping, the hardships and anguish of the dispossessed exceeded what was acknowledged in accounts written by the victors. Segments of some invaded populations may welcome new regimes, as Athenians mistakenly thought Sicilians would and, in recent memory, the Don Cheney and Donald Rumsfeld arm of the Bush administration said many Iraqis would in the 2003 invasion. That the invaded will be better off after the invasion isn't as common in prewar rhetoric as vilification of the targeted group, but in combination with it a welcoming contingent is often held up as an enticement.

Evangelism provides another motive in some instances, and in extremes genocide takes care of the resident population. That the holy element is a national concoction is shown by the different deities cited in different epochs and places, and by the atomic matter history of the universe. A welcoming population has a point when invasion doubles as a liberation, as it did in the Allied liberation of Europe from the fascist Nazi occupation of World War II. In other cases, a new regime may be as bad as the previous one, the Japanese replacing the British in Burma, for instance, and eastern Europe in the wake of World War II going from fascism to communism, both in a world conquest mode. An invasion can also turn out to be less harsh than feared, as was the case in ancient Israel in the Persian era under Cyrus (6th century) compared to others in the succession of ancient Israel's oppressors. Where Babylon transported Israelites who might rebel, Persia under Cyrus returned them to their homeland.

In the European colonizing of North America, selective portraits of savages and the anticipated welcoming of the invaders were subordinate to what was widely regarded as progress in agriculture, trade, and industry, plus the literary, communications, and civil order of a more developed civilization. The thinking was that a more advanced civilization will prevail no matter what - and it usually does, thanks to numbers, weapons, and central administrations. Refugees from 1947 through the 1950s fleeing east Pakistan and Bangladesh were another matter; not a transition from backward civilizations to modern cities but more a matter of nationalism, ethnicity, and ongoing conflict between Muslims and Hindus. Over half a million fled during the 1965 India/Pakistan war. In the Bangladesh Liberation War of 1971, several million East Bengalis left for India. Numbers have approached that elsewhere, most recently in Syria, where flimsy temporary dwellings have stretched for miles in the desert. Foreign countries have had to look for ways to accommodate over 5 million registered refugees, the doing less of doctrine and ideology in that case than of an intolerable strongman rule. More localized and temporary evacuations attend wildfires and floods, the story in 2019- 2020 Australia and other areas affected by global warming. Figures as high as 300 million have been proposed for lowland evacuation if the seas rise as much as they can and, as currently warming waters under Antarctic ice indicate, probably will. The Helots of Sparta and felons shipped from England to Australia are other noteworthy examples in the past. In any such episode, ancient or recent, the loss of a home place brings prolonged isolation from sources of information and schooling.

Nothing in the prospective 21st century suggests an end to such episodes. Likely to decrease if not end altogether, however, are populations driven out or taken over in large- scale conquests. Contested territories like Palestine and confrontations like the one between Russian-oriented and European-oriented factions in the Ukraine are too ingrained to disappear overnight, but they aren't likely to bring overrun areas comparable to those of the empires and colonies. Invasions by European powers in the colonial era will turn out to have been a one-time affair based on the technology of the rifle, wind and diesel powered ships, and the prolonged global replacement of foraging with farming and industry (a major factor in the colonizing of the Americas). Nothing further of that kind seems likely in an age of missiles, radar, nuclear threats, high altitude surveillance, and dense populations that can't disperse. Terrorist groups like Al Qaeda and Isis have conquest ambitions, but no chance to match an Alexander or a Genghis Kahn. Rather than having an air force, army, and navy, terrorists (with a few exceptions) have used homemade bombs and automatic weapons, and attacked selected targets rather than nations and entire regional populations.

Mastering advanced weaponry requires science and technology, and these aren't comfortable with the misinformation and devotion to the improbable earth history that has characterized religious beliefs since they started being recorded. It was mainly outside military help from less radicalized places that dislodged Isis from Syria in 2019. Yuval Noah Harari in the NY Times (May 24, 2019) put the limits of terrorist fanaticism outside its own homeland areas nicely, discussing the 9/11 attack and the capacity of humans as opposed to normal beasts to believe nearly anything: "No rabbit would have been willing to crash an airplane into the World Trade Center in hope of being rewarded with 72 virgin rabbits in the afterlife." No widespread conquest campaign will be so based either, only feverish movements organized by groupthink dogma. This isn't to minimize the shock to those who get invaded, driven out, blown up, or gunned down, merely to limit the likely scale of the fanaticism pitted against the strength of established nations. What the beliefs of the faithful amount to is best put in context not by a rival faith, but by the natural continuum and based on inductive observation.

What Imaginary Homelands Exclude

In what is imagined not as a displacement but an ultimate homecoming, like Dante's pilgrims, the future residents of an alternative place are urged to dissolve or lessen other commitments ahead of time. The prime source of the Christian version of that is Jesus arrogantly telling his listeners, "He that loveth father or mother more than me is not worthy of me: and he that loveth son or daughter more than me is not worthy of me" (*Matthew* 10.37-38). That prepares followers for leaving home in the usual sense to go Home in a transcendent sense. The landscapes for such paradises aren't usually particularized, but apparently not many of earth's species will make the transit, and none of the biological and botanical vulnerabilities. Faith in such an eventual homecoming has often been encouraged by regimes that aren't inclined to improve their own realms. Let death, not the current social order, be the liberator and equalizer.

In less radical displacements, something special is selected for transfer. Digestive systems, arterial distribution of nourishment, and much of the brain can be discarded, since their function is to cope with material things native to the planet and present in animal chemistry. Whether the final place is a golden city or an outdoor paradise, it doesn't have the private living quarters of castles, apartments, hovels, or houses, nor political subdivisions. These would continue loyalties and partitions that detract from the beatific vision. The alternative universe doesn't need fields of grain to feed the multitudes Captain Stormfield finds there, since nutritional and digestive systems aren't needed any longer. Other kinds of partition and isolation are also left behind, such as language differences, race, and, in some angelic versions, gender. Scholarship and archived knowledge, laboriously gathered information, and science aren't necessary where knowledge comes by revelation. Intellect becomes marginal, if not altogether obsolete.

Those who endorse one concept of such an idealized realm often set themselves against those who endorse another, thus Ambrose Bierce in *The*

Devil's Dictionary defining *Elysium* as "An imaginary delightful country which the ancients foolishly believed to be inhabited by the spirits of the good. This ridiculous and mischievous fable was swept off the face of the earth by the early Christians […] Nothing could be more

ludicrous than this crude conception; instead of golden clouds, harps, crows and a great white throne, there were fields, groves, streams, flowers and temples." Bierce, too, takes no notice of the population problem, as severe for Elysium as for the conventional place of harps and hymns. A secular utopia doesn't need a miraculous restoration, but it may require an extended voyage to get some distance from any of the known social orders.

Although the improved places of the utopias make fewer improvements on nature, in them too family residences that divide people give way to enclaves that join them. Tribe, race, and gender are subordinate to the utopian ideal. The focus on reformed politics would logically eliminate the administrative machinery that comes with hospitals, transportation systems, waste disposal, and much of the rest of a standard social and economic machinery, where physics, chemistry, and biology still reign. Storms and wildfires don't rage through utopias, floods don't overwhelm, and armies don't march.

Crime and policing are reduced, or disappear altogether. How much of life on the planet is difficult to handle is clear in how much such places exclude. As usual in human improvements of roughhewn nature, the universe needs a lot of work to make it less crude and hostile to life.

Back on the planet - also a home where places exclude a good deal. Within civil orders, what is allowed and what is forbidden by way of intrusions, theft, and scavenging is a matter for legal systems to define. With copyright and plagiarism, the waters get murky, since nearly everything people say and do comes from prior examples. Modern electronic intrusion and borrowing come more quietly and through more means than in the days of castle lookouts and spies. When they are secretive, however, they still amount to Polonius behind the curtain. To place pop-ups that interrupt internet searches and ads that crowd in into a brick-and-mortar setting, it is as if a salesman from down the street bursting in to pull away someone else's customer. In the transactions we know best, parasites are among the invaders. A new one appears whenever something uninvited appears on a monitor. In legislatures, unpopular riders that wouldn't pass on their own are tacked onto measures that will. Billboards line highways. Product endorsements become prominent on articles of clothing, especially the sneakers of athletes. Attention-getting product promotions include automated phone calls, information leaks, data theft, and surreptitious sales of private information of the kind that recently landed Facebook in trouble. In dystopias, such intrusions are more likely to be collected into a single dictatorial figure and his minions down through the bureaucracy.

Breaking In: Surveillance and Intrusions

In most real-world situations, intrusions of many kinds are forbidden by law - commercial soliciting by telephone, for instance, door-to-door peddling in some places, and anything excessively loud. Despite that, so frequent have intrusions become in the age of digital devices that telephone companies and gadget makers have added caller ID, call-blocking devices, and malware removers. These aren't completely effective and don't protect against every known kind of intrusion, information leakage and pilferage. That the National Security Agency (NSA) siphons private information became public knowledge with Wikileaks (2006) and the Edward Snowden affair (June, 2013).

Credit-rating companies gather private data and sell it or make it public. Early in 2014, President Obama addressed government sponsored information gathering as a policy matter and gave assurances of civil liberties, but these didn't extend to suspects imprisoned outside the country and being forcefully interrogated. Nor do the protections currently extend to children held in internment camps separated from their parents or to privately sponsored pilferage, 'cookies' planted in computers, and other forms of intrusion and eavesdropping.

As the capacity and speed of the devices increases, so does the craft of intrusion and theft. The raised volume of speech from a podium and, at one time, the turned-up volume of ads on radio and television are the publicity sides of that. University courses are now offered on how ideas go viral and on the "science of social epidemics." In their way, these can be as pernicious as real viruses such as Covid-19. Intangible things, like ideas, and small ones, like viruses, have no difficulty getting into households. Landmark studies of discourse such as Michel Foucault's *Archaeology of Knowledge* (1982) and M. M. Bakhtin's *Dialogic Imagination* (1982) neglected the viral attention-demanding, pollution, and pilferage aspects of discourse, but in *Discourse and Social Change* (1992) Norman Fairclough's survey includes them under power relations. Every sort of hustle that once stood behind podiums or on stumps and steps is now more likely to sit before a camera or at a studio microphone. Broadcasting expanded exponentially with the need for markets that came with the industrial revolution, and with the circulation of funds from companies to workers and to the commercial outlets they use.

Legal procedures in democracies ordinarily guard against certain kinds of forced entry such as arbitrary vehicle stops, searches, and arrests. Nothing equivalent protects against unannounced surveillance committed in the name of national security. Insofar as discovering real dangers benefits citizens, that is for the good of the ones spied upon - Big Brother does sometimes have a point. Brier Dudley's blog in the *Seattle Times* (December 11, 2013) advises, with tongue in cheek, that information gained from listen-ins could be put to better uses, such as tracking customers in department stores and directing them to what they might want or pulling personal files when someone enters the DMV: scan the face, pull the file. The foal, lamb, and infant at its mother's breast are life-sustaining examples of other intrusions. Among the more desirable invasions and extractions is the surgical removal of an unwanted growth - the sharp edge intrudes and out comes the growth. The police arrive and the burglar is handcuffed and hauled off.

Imprisonment without trial in some places in the 20th and 21st centuries has regressed to the arbitrary legal systems of pre-revolutionary Europe, when in the Puritan rebellion timely fair trial in England was one of the unmet demands of Diggers and Levelers, among the homeless of their day. Unjust seizure and imprisonment have sometimes fostered rebellions, as they did in Apartheid South Africa and in the Americna civil rights movement of the '60s and '70s. They continue to characterize dictatorships and extreme right-wing political movements. The novels of John le Carré and Alan Furst would have much less to work with but for secretive infiltration, news manipulation, and imprisonment without due process. A good many doors were in fact knocked on in Nazi Germany and the Soviet Union in the mid-20th century and those, along with China's wholesale conversion from one regime to another, are only the notorious cases. Exceptions to common rules against spying get more frequent the more threatened an authority thinks it is. By way of routine policing, doors can get forced even in democracies, if in nothing like the numbers under strong-arm regimes.

Nonviolent encroachment in computer invasion graduated into an international affair in Russian intrusion into the American 2016 presidential election, and in President Trump's invitations to foreign powers to do the same again in the 2020 election. None of these incursions, however, begins to match that of an actual military invasion. Bombs and missiles are dramatic breaking and entering devices. Radical political change is partly what brings about draconian methods, though the abuse of power is multi-faceted, age-old, and needs no special causes. For some forced entries we have defenses, for every sword and spear a shield. Needless to say, these aren't completely effective. As television and book portrayals of KGB, SS, FBI, CIA, NSA, MI5 and MI6 show, such agencies ignore common decency rules of surveillance and spying. Under capitalism and parliamentary democracy, the knock on the door comes in daylight hours and most likely isn't from a government agent but from a salesman or someone fulfilling an evangelical mission. Out on the street, the demand for attention comes equipped with loud speakers, banners, murals, placards, billboard slogans, organized marches, and rallies. Patriotic observances intrude in all manner of gatherings from convocations to sporting events.

That they are unifying nationally is obvious. That's the purpose of national anthems sung before an event unrelated to nationalism. That they increase tensions and raise trouble internationally is almost as obvious.

Because powers added to executive authority are seldom rescinded, we can expect legalized breaking and entering to continue along with a commercial barrage of attention- demanding salesmanship. Imaginary or real, unseen monitors aren't assumed to be everywhere - except where religions claim the omniscient oversight of a supreme being. In contrast, for a motorist guarding against getting a traffic ticket, the suspected motorcycle cop is only behind the next billboard, not in the cornfield alongside the county road or up in the sky with an unobstructed view. Signs warning of radar are designed to spread wariness over a wider area than can actually be patrolled. Surveillance is an instrument of privacy as well as of privacy invasion. Stores, hospitals, apartment buildings, and private homes set up cameras in strategic places activated by movement. To discourage breaking and entering, they post notices of their devices.

Monitoring becomes ticklish when ethics says one thing and authority something else. Whistle-blowers are usually not at ease with themselves and onlookers not at ease with them. Intimidated by threats and peer judgment, even eye witnesses to crimes sometimes keep silent. In cases of sexual harassment, a very personal form of invasion, victims more often than not remained silent before 2017, when collective support encouraged them to come forward, as a good many have since done. Alertness to incursions is usually stationed at national boundaries, but equivalents come on other levels and in other places. The spider in its crevice, the prairie dog colony, and the gopher in its hole are like sentries on the ramparts on the lookout, and the reason is similar: getting information in a timely manner and seeing before being seen are survival advantages. That's true at many levels, from snails whose horns pull in to national early-warning systems equipped with radar, spy planes, and satellite transmission.

Relics As Information in Decay

Information streaming into a home place comes not only through what arrives in print and through the air, but in what a house and grounds have lying about or stored away. Relics come in all ages, from last year's discards to millions of years represented in species and billions in rock formations. One of McLaurin's settings, in addition, had relatively ancient stone age implements and tools. If anyone put these and recent discards together in a sequence, the collection would span several thousands of years - not a great deal bu,t together with the geology, biology, and botany, a good stretch of the continuum. In his case, which I'll put to use again, a stone implement that looked like a pestle had probably come from a nearby knoll known as the Indian mound, a name carried over from the first wagon train homesteaders over a century ago. The artifacts that indigenous people had left behind had been in use from the time they arrived several millennia ago to the early 20th century, and some of them, no doubt, had prolonged Asian and African histories. In any case, similar artifacts had originated in Africa and traveled through Eurasia long before human migrations reached the Americas. The only crafted items as archaic as the pestle were two crude stone arrowheads. In a small fraction of the time that primitive tribes had been there, the rusted parts of an old tractor in the barnyard had been listed in parts catalogues. Where bones appeared among the leavings, they revealed diet to paleontologists and that, in turn, told of ways of life. Nothing that ancient was included in the local relics, but *National Geographics* stored in the attic of a century old wood frame house had accounts of them.

The more recent relics made of iron and wood were in better condition and put advances in machine technology on display. When compared to the older relics, they underscored the past century's accelerated invention as well as the change in mode of production nearly everywhere, from the hand-crafted tools and utensils of foragers to mechanized agriculture. More devices had gotten designed and distributed in the past few decades than over the past several millennia. In some two centuries, locally and in many other places, technology had gone from drum beats and smoke signals to newspapers, radio, and electronics. The more recent times of the later generations of European and American immigrants were represented not only in junkyard discards but in an ancient barn still in use. That no houses within several miles had survived more than a century could be attributed to their lack of usefulness as animal shelters. Their sites were needed, and so they had been torn down or burned. Only a few here and there survived. In that barn built a century ago, a doubletree, horse collar, and other items from the workhorse era lay under a battered L-shaped workbench in a corner. The shingled roof overhead, though leaky despite its corrugated metal patches, was delaying their blending into the soil. A long unused harness hung from nails on a rough board wall. The discarded machinery parts came in John Deere green, Allis-Chalmers orange, and Gleaner unpainted aluminum. The hub and iron rim of a wagon wheel were missing half their wooden spokes. In the barnyard under an oak tree was an iron-wheeled hay rake with an iron seat and a lever used to raise and lower curved steel tines. Horses had pulled it initially and later a lug-wheeled tractor. Later still, a small rubber tire tractor. In an oak grove above the barn, a rusted tractor with flat tires kept company with a plow, a disk, a drill, and an abandoned truck. These had been turned over to rust for the final stage of their breakdown. A good deal of the story of the global transition from tribal wanderers to property-owning growers showed in the difference between that elaborate machinery and the stone age artifacts. Locally, that story was only two centuries old; elsewhere, it went back much further.

Animals had left what animals leave as well - mainly bones. Droppings from sheep that had departed decades ago crusted a section of the barn where milk cows, too, were once stanchioned. Their absence told of the coyotes that had ended local wool and mutton production, one of the determined responses of the wild to human encroachments. This had been coyote, bear, and puma territory for millions of years before people arrived. As the weedy pastures also showed, domesticated plants were as helpless among wild ones as dogs among wolves. Other domesticated animals had also left: saddle horses for parades and rodeos, cattle for industrial level beef production, laying hens for cages lined in rows, dairy herds for farm factory milk sheds, and hogs for indoor cages in long feeding sheds. Rodents and a nest of barn owls remained. Would an anthropologist a few centuries hence be able to trace the progression step by step? Lacking other records, probably not. Distance foreshortened and compressed intervals and lost the logic of a given succession. In the enactment, intervals came in small increments; in the reconstruction, in centuries.

Would anyone be able to guess the function of the oversize iron fork used to lift loose hay from a ground level wagon to the loft? Looking as alien as the helmet and visor of an ancient warrior, for over a century it had dangled from an overhanging metal track. Suspended from a heavy hay rope threaded through a pulley, the other end attached to a tractor hitch, this odd contraption had drawn hay up to a carriage running under the roof along a metal rail. When it clicked into place, the hay jerked inward like a great wig snagged by the wind. Released by a trip rope, it landed on the hay mow floor in a cloud of dust. The oversized device then reversed its journey. When hay bailing came in, that kind of storage had become as obsolete as the wood-domed radio in the basement of the house.

Altogether, the relics told of an economic and cultural make-over that, in about a century, had revolutionized food production in time for the post World War II population boom. Although not a poet, McLaurin found verse an apt way to pile layers atop one another as this place had. Inventories, histories, chronicles, diaries, and newspaper articles put a few representative samples in orderly storage. Lifting pieces of what had once been there line by line seemed more akin to the way memory collected its items and filed them away. His only venture in that mode came out less a narrative and even less an inventory than an assortment, not as scattered as the miscellany lying about but more disconnected than a chronicle would allow:

The seasons return in their circuits.
The long way north narrows to a rutted lane
ending in a deer trail
crowded by hawthorn and hazel.
Blackberry vines sprawl under ocean spray.
Wild roses stare at hillside stone
stamped with leaves and the bones of fish.
An aging barn leans toward repose
and lets board ramps down for mice.
On knees printed with woodgrain
I nailed this sagging hayloft floor
while an elder shingled overhead.
Through windripped gashes in the roof starlight comes from a ghostly past.
A long way south the same stars shine dimly through dense brown air
rising from freeways like river mist.

Putting eras and regions together first required setting boundaries as historians did; personal items went into diaries and photo albums, framed pictures went onto fireplace mantles, a few artifacts made their way into museums. To go deeper and more systematic required anthropologists and lineage services with DNA tracking capacity.

Deeper still fell to archaeology, paleontology, physics, and cosmology. Was that descending into history or rising up to it? Perhaps both simultaneously, like Escher stairwells or like telescopic deep scans of the stars going up and back. Spacetime itself had no up or down, only myriad locations.

When he looked to others to see how they went about connecting local places to past intervals and settings, he found poets prominent among them. Every method of recording had its own way with relics and the remembrance of things past. A poet's was likely to be self-oriented and easier to decode than those that made their way into anthropology.

Records of environmental shifts, like the ice ages, were the sources of simultaneous species extinctions and origins. Chemistry and physics drew the most exacting jobs - the decay rate of lethal protactinium, for instance, carried from thorium to helium in the range of 32,000 years. Neanderthals were still roaming Europe that long ago. In "The Poem That Took the Place of a Mountain" (1952), Stevens' simulated search for home ground coincided with one for words that could connect here and now to the age of a mountain:

> There it was, word for word,
> The poem that took the place of a mountain.
>
> He breathed its oxygen,
> Even when the book lay turned in the dust of his table.
>
> It reminded him how he had needed A place to go to in his own direction,
>
> How he had decomposed the pines,
> Shifted the rocks and picked his way among clouds,
>
> For the outlook that would be right,
> Where he would be complete in an unexplained completion:
>
> The exact rock where his inexactnesses
> Would discover, at last, the view toward which they had edged,
>
> Where he could lie and, gazing down at the sea, Recognize his unique and solitary home.
> (*The Palm at the End of the Mind*, 374)

As again in "The Planet on the Table" (1953), the elegiac side of that was the unique and solitary part. True, Stevens hoped to gather a community of like-minded readers, but it was up to them to come through the opening he provided, and he wouldn't personally be standing in the doorway to greet them. With his collected works shelved in private and public libraries, he would become something of a relic - but then, so were Homer and Shakespeare.

For the homesteaders of the place with the castoffs, building a durable place had been quite literal. As their journals and diaries indicated, their early years on land-grant acreage had been demanding and labor intensive. They were doers, not poets and historians. Taking over land that had never been tilled or pastured required brush clearing, fence building, planting, house and barn construction, furnishing, and decorating. In a geologic timeframe, it was over in a flash; even in human migration terms, it was only one chapter in a long history. To them, however, it was a matter of daily persistence that consumed their entire adult lives. The first few generations of Territory pioneers had seen little technological advance. A journal left in an attic notebook filled in a few details of that stage. *A habitation must be secured and the only way to secure it,* written in an elegant hand meant to be readable for centuries, *was to cut the trees and build.* The tree cutting had been by axe and crosscut saw, examples of which were among the castoffs in the barn (one rusted saw and one double-bladed axe lacking a handle). Instead of spikes and nails, wooden pegs had held the first buildings together. Homesteaders in the Territory had turned to hunting and gathering to supplement their diet, but the venison, rabbit, trout, and wild berries were no longer as plentiful as the first migrants had found them. As the clearing and planting progressed, those former Asians had quietly retreated into the coastal hills and upstream along the rivers, occasionally coming to farmhouse kitchen doors in need of handouts. With hunting and gathering's produce limited, they looked hungry.

Many of the records were no longer kept at home but in county court houses. Before tribal territory could become fenced property, homesteads had to be surveyed and property lines recorded in degrees, minutes, and seconds. Places were assigned several specifications besides location such as soil types, successively deeded ownership, and tax rolls. Nearly everything was registered - machinery, residents in the census, members of graduating classes, social security contributors and receivers, and burial sites. Thanks to that record keeping, what could be looked up and retrieved from the past century had become voluminous. In the notation systems, what had been scattered in the enactment became better arranged, though a good deal vanished without a trace. The current residents received information less from the past than currently from around the globe.

Satellites relayed signals to a disk on the roof so that what was happening in Syria could be known in a few minutes. Anyone with something to say about that could get on line and say it.

No longer a resident anywhere close by, McLaurin was interested mainly in the deeper layers represented by bones and terrain. He couldn't interpret these well himself, but he knew where to find those who could. Meteors that had reached the ground carried messages from far out and long past. As celestial relics, they intercepted other sources of information at hand. None of the bits and pieces broken loose from their former moorings were meant to be information any more than discarded light and heat from the sun were meant to become leaves of grass. Scavenging and collecting made the conversion. McLaurin was getting better at that, and took pleasure in tugging on this or that to see what came with it. Doing so converted relics into artifacts and settings into museums, not altogether collecting and preserving of information. Should he need them, those greater depths of time and space were always there as information in electronic and printed storage.

CHAPTER TWELVE

DON'T ENTIRELY CURB YOUR ENTHUSIASM

The Argument

The planet may have the resources in its biosphere for its current level of population to keep doing much of what it is doing, but atmosphere, fresh water, and food demand more coordinated attention than the disunited nations are accustomed to giving them. Correcting world views incompatible with natural history and with one another would lower the intensity of several major current distractions,

< >

Too often, politics and politicians have not served us well on environmental issues. . . . Thirty-second television commercials and sophisticated public opinion polling can now calibrate and target a political message with frightening speed and accuracy, and they can do more to manipulate the opinions of voters in two weeks than all the speeches and debates and political organizations together can accomplish in ten years. (Al Gore, *Earth.*)

< >

What We Need to Know and Why We Need to Know It

Notions incompatible with atomic matter history prevail as much today as they did when Carl Sagan lamented in 1980 that "there are those who secretly deplore [...] great discoveries, who consider every step a demotion, who in their heart of hearts still pine for a universe whose center, focus and fulcrum is the Earth." Not just the earth, but mankind, one of many million species thereon. Such reality denial isn't as prevalent in the 21st century as it was several centuries ago, but it remains surprisingly entrenched. Sagan continues, "if we are to deal with the Cosmos we must first understand it [...] Understanding where we live is an essential precondition for improving the neighborhood" (193). Change *improving* to *salvaging* - so much have vital resources, especially moderate air temperature and fresh water, deteriorated over the past several decades. In the US, campaign contributions and gerrymandered voting districts give a disproportionate influence to anti-environmental interests and to nationalism. The latter can be a powerful instrument for organizing global efforts, but it can also be divisive enough to escalate policy differences into wars. Talk show hosts, one major television network, and one of only two influential political parties make fact twisting, outright denial, and trigger-happy nationalism almost standard procedure. Perversely, a cooperative, peaceful, habitable planet is low on the list of what conservative interests seek to conserve.

Take human contributions to the climate problem, now widely acknowledged and quite substantial. As is clear to those who get very far into it, absolute certainty in matters of climate isn't possible. High probability is, however. Why the areas of doubt? Airs swirls and water flows. Hence temperatures measured at one time and in one place represent only what is true now, this moment, in this place. Depths in water and altitudes in air come in layers that mix. Weather adds up to climate only in increments. March 2019 in the northern US was colder than usual and nowhere on the North American continent was warmer except Florida. How is that to be weighed against other regions over decades? Over centuries? You see the problem. Overall, though, there isn't a reasonable doubt, and the chorus of voices saying so has thankfully grown over the past few decades. CO_2 (carbon dioxide) unquestionably contributes to global warming, and CH_4 (methane) is a potent threat. The worst possible consequences of these two, combined with several other greenhouse gases, can't be ruled out. Those consequences include not only coastal and lowland evacuations but quite possibly mass extinctions.

Almost everyone who has looked very far into climate with an open mind acknowledges that. Still, regions in the US disproportionately powerful in the Electoral College remain careless and toxic in influence.

Robert Bryce (2008) is typical of fossil fuel spokesmen in finding oil "a strategic commodity. And it always will be" (15). He is, of course, correct to note that "regardless of the ongoing fears about oil shortages, global warming, conflict in the Persian Gulf and terrorism, the plain, unavoidable truth is that the US, along with nearly every other country on the planet, is married to fossil fuels" (8). That is an accomplished fact.

Considering the gravity of the problems, however, the dismissal of shortages, global warming, recurring wars in the Middle East, and terrorism in "regardless" is nonetheless cavalier, and "always will be" is false if one takes *always* literally. Oil doesn't come in inexhaustible quantities. Marriage is subject to divorce, and in this case the separation will come.

The question is whether or not it will be soon enough to avoid the tipping point of greenhouse gas release in which this year's excess gets added to next year's, and compounded they increase the next, and so on to a runaway point. The seas have been warming, though not irreversibly so far thanks to the high energy required to raise the temperature of water. One of the exceptions is the 2020 discovery that water under Antarctic ice is warming more than predicted. The melting of polar ice and permafrost makes it already doubtful that disastrously rising seas can be avoided. The question is how much further they will rise. Displaced lowland dwellers in high numbers can be counted on to disrupt civil orders, and that will direct attention and resources to urgent problems of people at the expense of attention to what is bringing them about.

Explaining the basics of natural history to those who will inherit the planet will affect how they get around, where they live, how many children they have, how they heat and cool their dwellings, who gets their vote, and how they apply critical reason to specious ideas. Even some potential terrorist recruits who aren't hopelessly lost to reason can see that scuffling on the bridge while the ship heads onto the shoals isn't wise. Removing motives for ethnic and national intolerance comes cheaper than removing combatants, and it leaves healthier minds in still living bodies. The problem isn't limited to fanatics; generally credulous populations help extremists go a step further. That may not be provable, in fact it probably isn't, but how the stadium effect works suggests it under the dynamics of groupthink, and wars based on nationalism, religion, and ideology testify to it. In any case, if *this* ship sinks the passengers will have nowhere to go. That takes precedence over nationality, gender, race, ideology, ethnicity, and any other principle of people-clustering.

Since the problem in large part is what and how people think, the only fix is education. Our grouping instinct binds us to clans, ethnic groups, and nations, not to those who know such things as the periodic table and what brought on previous extinction episodes. For the billion people in the world who can't read and don't have classrooms with visual aids, and for the fifth of Americans who can read but not well, non-specialist vocabulary and visual displays will serve well enough – and they will have to. That comes preliminary to any fix that is broadly based and lasting. The fundamentals of natural history can be simplified without doing serious damage to them. It is these, and not any particular slant on them, that make storybook substitutions implausible.

Not only formal education, but also the media can bring a better understanding of the climate problem and the natural continuum. To their credit, they have been seriously at work on it for some time. The November

2015 *National Geographic* headed *Cool It,* for instance, was devoted to the repercussions of climate change and fixes for it. Given the magazine's reputation and circulation, it undoubtedly had a positive effect. A recent *National Geographic* world atlas devotes several fact-laden pages, maps, and color-coded graphs to regional changes recorded over the past few decades. The school systems have also not been negligent, as the vocal presence of young voices has been demonstrating over the past few years. This upsurge in 'habitable planet' activity would grow and be better armed if natural history overall was more broadly known. The extremely narrow span of living conditions would then become clear, and the chances of data deniers getting elected or appointed to office would shrink.

Nearly a century and a half after Humboldt's *Cosmos* (1858), Edward O. Wilson put the natural history overview this way in *Consilience* (1998): "The ongoing fragmentation of knowledge and resulting chaos in philosophy are not reflections of the real world but artifacts of scholarship" (8). That people tend to see reality "through the lens of ideologies and religious dogmas or commanded by myopic responses to immediate need" (13) remains as true now as it was when Wilson, like Sagan before him, credited sectarian partisanship for much of science denial: "Most of the issues that vex humanity daily–ethnic conflict, arms escalation, overpopulation, abortion, environment, endemic poverty, to cite several most persistently before us–cannot be solved without integrating knowledge from the natural sciences with that of the social sciences and humanities" (13). This has been my contention as well, and it isn't a new idea. The age-old conflict between entrenched beliefs and observation-based common sense and theory that resumed with Copernicus, Kepler, Brahe, and Galileo came to a crisis in Darwin. It continues in fact deniers holding hands and skipping toward the cliff like carefree lemmings. The fundamentals of natural history aren't debatable, aren't avoidable, and aren't kind and loving.

Those who are already convinced that the atmosphere is a crucial part of the crisis matrix, that shortage of drinking water presents a serious problem, and that population is a serious matter of concern – who don't need further facts and figures – can if they want proceed to "Affordable Change." The following section condenses what a consortium of disciplines have put on record over the past few decades. Though data continues to accumulate, for some time it has been more than sufficient to show that life on the planet is seriously threatened and falls, to a considerable extent, under human influence. A mechanized human population of 7 billion has been doing serious damage to the planet, especially to its air envelope. At the forefront of what can go wrong is the reaching of a tipping point in methane release. We have the example of planet Venus to tell us what happens with high greenhouse gas concentrations, and we have the Permian die off and several other extinction episodes to show what happens when nature alters course.

Some Critical Data

Websites with charts, facts, figures, and explanations concerning the natural continuum and climate change in particular can best be accessed simply by browsing for them. Doing so produces the findings of the World Open Data Catalogue, the National Oceanic and Atmosphere Administration (NOAA), the Intergovernmental Panel on Climate Change (IPCC), NASA, and other agencies which have compiled data and put together charts that show the extent of increases in greenhouse gases over the past few centuries. A sharp upward curve projected for the rest of the 21st century was behind a recent grim UN warning based on revised estimates. The costs in terms of wild fires alone in the American West and Australia have run in the many billions, inflicting suffering on millions, killing quite a few, and killing animals in the billions.

The IPCC warns policymakers that the "warming of the climate system is unequivocal, and since the 1950s, many of the observed changes are unprecedented over decades to millennia." It cites measurements of carbon dioxide, methane, and nitrous oxide (N_2O), concentrations of which "have increased to levels unprecedented in at least the last 800,000 years" (cited from "B.5, Carbon and Other Biogeochemical Cycles," available online at climatechange2013.org). That estimate is in accord with the 2010 EPA report likewise available online. US carbon emissions for 2018 rose some 3.4%, despite coal use playing no part in the increase. Contrary opinions aren't difficult to find, but data compiled region by region over time, around the world, plus ice samples for years in the hundreds of thousands, leave no reasonable doubt of a significant increase in greenhouse gases and their effect on climate.

One recurrent argument against taking climate change seriously has derived from a temporary lull in the rate beginning in 1997 which initially sounded more comforting than it has turned out to be. The "case of the missing heat" as *Nature* called it, along with other short-term climate oscillation, isn't puzzling in the long view. The publication of the global heat chart for 2015 shows a predominance of areas of increase. The following year looks no better, the third in a row that global temperatures have set records. January 2017 was the third warmest January in 137 years, NASA's Goddard Institute for Space Studies reports, and so it goes into the 2020s. Rising and lowering warm and cold currents in the Pacific are tricky and result in complex computer models of Pacific Decadal Oscillation

(PDO), but atmospheric temperatures do respond to surface sea temperatures and vice versa. Crediting warmer seas for hurricanes stronger than usual in the Caribbean, Texas, Florida, and up the east coast is somewhat hypothetical, but sound enough. Climate change skeptics sometimes cite water vapor as an important influence on greenhouse gas, and indeed it is, but it doesn't reduce the importance of substances under human control. A prolonged global cloud cover isn't anything future generations can count on.

The reasons for concern about methane specifically are its initial strong effect, its lasting power, and its abundance. Its average lifetime in the atmosphere is estimated at about 12 years, compared to nitrous oxide at 121 years and fluorinated gases at a few weeks to thousands of years. The New Zealand contributor to Wikipedia Alan Liefting warns that, over the long run, methane's potency computes to a warming potential 34 times that of CO_2. That CO_2 also contributes to the planet's green growth is less encouraging than it might seem, since overnight 'respiration' returns to the atmosphere some of the what the new growth has absorbed. Charts and graphs of regional methane release dating back to 1740 are among the supplementary materials of the IPCC report *Climate Change 2013.* Also available to science watchers is a Cornell University study of methane release from hydrofracking (2012). Add to carbon dioxide and methane HFC, HCFC, N_2O, propellants, refrigerants, and nitrous oxide. These aren't high in amounts compared to carbon dioxide and methane, but in collaboration they do contribute substantially to atmospheric warming.

Thanks to hydrate concentrations in the oceans, the former EPA, before it was crippled by an administration dedicated to fossil fuel, found that they would "eventually be affected by climate change." Since emissions from the seas naturally increase as air temperatures do, "temperature changes are likely to be far more significant than changes in pressure" that might constrict them ("Gas Hydrates," Chapter 7 of the 2010 EPA report reiterated in 2012 and 2014). The rate of human generated emissions in the Executive Summary of the EPA report is higher, at 60%, than natural emissions. The former breaks down into landfills 18%, coal mining 10%, manure management 10%, natural gas and petroleum 29%, and enteric fermentation 26%. Overall estimates set harmful emissions from natural sources at 208 teragrams per year and anthropogenic emissions at 396 (13). Marginally lower is the IPCC estimate of human controlled releases of roughly 330 teragrams from landfills, livestock, fossil fuel consumption, and biomass burning. Gas emissions from livestock are more serious than schoolyard humor implies. Nature's release from wetlands, forest fires, plants, termites, and oceans runs around 270 teragrams, or a split of about 45% natural to 55% human.

In one well documented estimate, critical gas capture falls about 7% short of absorbing what escapes. Possibly as a result, in a moderately quick response, the years 1999-2007 saw a record thawing of permafrost in Alaska and Siberia, and in 2019 an unexpectedly sharp rise in atmospheric methane plus another increase in melting permafrost. One of several internet articles calculates that melting yedoma, "a type of permafrost," comes to 4 teragrams per year. The "possible adverse effects projected as gas escapes [...] are estimated to have the potential of a sixty-trillion-dollar impact on the world economy." That is a necessarily rough estimate, but an already significant financial impact of greenhouse gas buildup over several decades is computable in days of labor lost and in deaths. The latter from air pollution, much of it from coal consumption, are estimated in India at 100,000 a year and in China at 750,000 - far more than the fatalities from the Covid-19 viral pandemic in 2020, but without the emergency response or the publicity. Health isn't exclusively a foreign problem. In the estimate of the Clean Air Task Force drawing again on the former EPA, some 14,000 to 36,000 people die annually in the US from pollution, and survivors lose over 1,600,000 workdays. For that and other reasons, the Clean Air Task Force report in September 2010 concluded that strong regional caps on fine particulates (SO_2) were cost effective. With coal industry leaders already of the opinion that jobs will not be returning to coal that may be a mute issue.

Though particulates are different from greenhouse gas emissions, so is anything lungs handle well, and they combine in the burning of coal and in auto emissions.

Affordable Change

On the better safe than sorry principle, I agreed earlier with Lester R. Brown's (2008) plea to treat the avoidance of a tipping point in greenhouse gas release with wartime urgency. It took the bombing of Pearl Harbor to turn the US from isolationism to its World War II effort, and pictures of stranded polar bears aren't as adrenalin producing as those of bombed and burning battleships. Greenhouse gases are invisible. Rising ocean temperatures aren't noisy. Rather than going into high gear, the US at the federal government level has grown sluggish. Several privately funded institutions have had better foresight than the public sector recently (2020), hampered as it is by the need for political candidates to solicit election campaign contributions. Among those in the forefront has been the Rocky Mountain Institute's "speeding of the energy revolution," capitalizing on the work of Paul Hawken, Amory Lovins, and L. Hunter Lovins in *Natural Capitalism: Creating the Next Industrial Revolution* (1999). The book's areas of concern include much of modern industry, building, transportation, water management, "food for life," finance, and departments concerned with public health. The charts of the World Bank Open Data Catalogue (datacatalogue. worldbank.org) show the scope of these and similar efforts to improve living conditions worldwide. They struggle against population increase, ignorance, indifference, and chronic water and air polluters.

That a conversion to clean energy can damage an economy is true of some sectors and not of others, but, in any case, changing energy sources needn't change the products. Going hydrogen or electric in powering cars replaces the fuel and engine type, but not axles, wheels, and performance. As to comparative energy costs, those of coal and oil are often set unrealistically low. In commenting on the 'I'll be gone' shortsighted syndrome, Thomas Friedman (2009) remarks that the real cost of fossil fuels isn't being calculated by either producers or consumers: "Every one of us who enjoyed using these underpriced carbon-emitting energy sources got to privatize the gains. The losses, however–the long-term impact of all this carbon building up in the atmosphere–we have socialized. We charged all of this on our children's Visa cards to be paid by them and their children's children far into the future" (23). According to the International Centre for Technology Assessment, reported in a geonomics "Progress Report" (20 April 2014), the true price of gasoline could run as high as $15.14 a gallon. Adding rising seas to the calculation raises the costs substantially if even just a small fraction of the cause is attributed to fossil fuel consumption.

Like the Worldwatch Institute, the 2012 *Scientific American* draws on a variety of studies that canvass resource problems rated as to effects and costs. Again, the list of repercussions linked to global warming is formidable, this time put metaphorically in terms of apocalyptic conquest, war, famine, and death. As to famine, Brown answers yes to the question "could food shortage bring down civilization." With the lowering water table and less irrigation, some regions have experienced a serious decline in water available for agriculture. Melting arctic ice may lower rainfall on the west coast, a large timber and farm goods area. Drought in recent years in China and in central California has affected a non-trivial amount of what Chinese and American people eat. The water table of California agricultural areas has dropped below what a few normal years of rainfall can restore. The rain and the snow pack of 2018-2019 were helpful, but not a fix for the prolonged effects of drought and the contaminated water of some 300 central valley communities. Water shortages are widespread and chronic in Africa and have reached a crisis in parts of Pakistan.

Critical points for these and other areas of the biosphere are nearing what Kaku considers the "most dangerous transition from primitive civilizations with their barbaric savagery to a civilization capable of space colonies" (347). Starvation and malnutrition have declined over the past few decades overall, but that is less comforting than it might seem when an estimated 805 million people lack enough food to maintain good health. The gains have been regional, with hunger decreasing in Latin America, the Caribbean, Asia, and the Pacific and increasing in Africa. Shifts in rain and snow amounts may help some areas, but populations live in afflicted ones, and food exports come from some of them. However tricky measuring global warming can be, temperatures recorded over extended periods from multiple locations are revealing. Microbes near the lower end of the scales could possibly influence future calculations, though that is speculative. Among the more curious of those that marginal living conditions can produce are the *methanomicrobia. Methanosarcina,* one of that group mentioned earlier, could conceivably become a WMD, again a speculative matter only, not a prediction. That member of the Euryarchaeota kingdom has been around since the Archean and, if a hypothesis proposed in 2014 about the Permian die-off turns out to be broadly accepted, we may have to charge it with being among the planet's greatest assassins. That prospect isn't critical to the concern over greenhouse gases, however, since they don't need microbial help to wreak havoc. The planet's most dangerous species at the moment is its most successful predator and inventor of mechanisms.

Alternatives

The planet receives from the sun more extractable energy than will ever be harnessed, and has additional supplies of its own that are clean in both production and use.

Extractable is the critical word; the energy stored in materials other than uranium, plutonium, and thorium isn't always economically feasible, though it has been from wood, coal, oil, natural gas, peat, and other combustible materials. Ash from fires hasn't. so far, been widely used, but it could conceivably become so. The more urban a civilization becomes, the more its waste piles up, and that adds UASB processing (upflow anaerobic sludge blanket) to the list of potential if minor sources of energy.

Without needing dams, several kinds of turbines harvest water power (a gravity production), the Tyson in midstream. Add tidal and wave power for coastal regions. Some chemical sources are renewable, and hydrogen as fuel can be extracted from several sources. Also marginal but with long-range potential is the harvesting of vortex energy from spinning air and water turned by hemispheric dynamics. I cite these only as reminders that, except to support the pertinent industries, we've no reason to stay addicted to fossil fuels even while they last.

Those who want an accelerated timetable in the conversion to a better combination of energy sources have a valid point in urging a prominent nuclear reactor option.

George Monbiot in *The Guardian* (2011) cites the Committee on Climate Change that opts for a 90% reduction of greenhouse emissions by 2030, faster than the US-China agreement and anything likely to grow out of the 2015 Paris conference. The goal is to limit global warming to 2^0 Celsius. While the fossil fuel gap can't be fully closed from renewable sources alone quickly enough, Integral Fast Reactors (IFRs) can make a significant difference while consuming already-existing radioactive waste (*The Guardian* website for 2 February 2012). As to timing, on behalf of IFRs Charles E. Till and Yoon Il Chang in *Plentiful Energy: The Story of the Integral Fast Reactor* (2011) warn that "The facts today lead directly to a . . . very simple conclusion. If we are to maintain our present prosperous manner of life, large amounts of nuclear-generated electricity will be essential. There is no other non-fossil source with the necessary magnitude. And this need is not far off. The outlook for fossil fuel sustainability is worse than commonly thought [...] The only result of this path, unaltered, is increasing shortage of energy–life-changing and draconian" (79, 82). Quantity and cost are indeed problems, but both fission and fusion can offer substantial help.

The history of nuclear power has admittedly been mixed, and the administrative tangles in promoting it can be formidable. Under the heading "America's Convoluted Energy Regulatory Structure" (323), Robert Bryce (2010) lists 12 US departments, a number of agencies, and over two dozen congressional committees with a hand in energy policy. They listen to nervous constituents concentrating more on the perils of atomic energy than on what happens if it is subtracted from the total. Richard Martin (212) adds that "the nuclear power industry in particular has been ruled for decades by technological lock-in–the tendency of established technologies to crowd other competing (and possibly superior) systems out of the market" (219).

Along with cost and technology, safety rises in priority in any proposal for more nuclear reactors. The Union of Concerned Scientists (UCS) concludes that only the "Evolutionary Power Reactor" looks significantly safer when the threat of terrorist intervention is added to that of meltdowns. Their report concentrates on the Nuclear Regulatory Commission and a limited range of reactor types currently in production or proposed. Where plutonium and uranium aren't involved, however, neither is their weapon grade improvement. In its 2007 report, the UCS finds the bottom line to be handling, processing, reprocessing, and transporting "large amounts of material that is usable in nuclear bombs," thus giving "terrorists numerous opportunities." That ignores thorium reactors and the progress being made on their development in several countries. Adding Liquid Fluoride Thorium Reactors (LFTRs or "lifters") strengthens the case for the safety of reactors that can be operational in time to make a difference.

Consider that source a moment; assessments of thorium as a reactor fuel aren't new, and are available to the public on websites and in book-length studies. The websites include discussions in the Energy Collective (theenergycollective.com), howstuffworks.com, and Wikipedia articles accessible to readers without advanced degrees in nuclear physics. Thorium reactor supporters – the nonprofit Thorium Energy Alliance, the International Thorium Energy Organization, the Russian Kurchatov Institute of the 1970s, and Flibe Energy – find affordable clean energy from that source within reach. As Barry W. Brook summarizes, "Imagine a reactor that passively responds to critical stressors of the kind that befell Three Mile Island, Chernobyl and Fukushima by shutting down without human operators even needed to intervene. Or one that includes a secure recycling and remote fabrication system that, almost Midas like, is able to turn uranium and even old 'nuclear waste' from contemporary reactors into an inexhaustible (and zero-carbon) fuel, as well as simultaneously solving the socio-political problem of long-term disposal" (*Plentiful Energy*, x).

Robert Hargraves offers similar testimony. Alvin Weinberg at Oak Ridge supervised the initial research, which Kirk Sorensen and others have dug out and used in attempts to procure funding. A new generation of physicists stands ready to do the supervising, some of them students of leaders in the atom splitting era of the 1950s-60s. The results of 1960s experiments at Oak Ridge got more or less buried in a 1972 shutdown, followed by a series of renewal attempts and further shutdowns. The original researchers gave up partly because of inconsistent government policies. The story of the new wave of scientists has been captured in a 44-minute YouTube presentation (youtube.com/- watch?y=knofNX7HCbg). Several articles and books have chronicled other chapters of the LFTR story, which has been unfolding over the past two decades. Lifters, their proponents assure us, can potentially consume radioactive waste that has already been stockpiled without incurring the risk of nuclear armaments or the meltdowns that have plagued aging uranium reactors. That appears to answer the concern of the Union of Concerned Scientists over waste that contains weapon grade plutonium and uranium.

Reductions in waste and a passive shutdown system would be significant advantages, as would the shortened half life of the waste that remains.

Housing the reactors in small areas is another advantage over fields of solar panels and wind turbines. Moreover, thorium is abundant. Its production and use are reasonably clean, and in mass production the cost is more affordable than realistic assessments of coal and oil production. The list of advantages is quite long if somewhat technical, the disadvantages brief and also technical. Several Research and Development projects are underway based on earlier Oak Ridge research, notably in Australia, Great Britain, China, Japan, and the US.

Where the Atomic Energy Commission and congressional bodies have been slow to move, some energy companies haven't. Transatomic Power and The Weinberg Foundation in Great Britain are current examples, the latter receiving the endorsement of the House of Lords in 2011. Comparable movements are underway in other countries less under the thumb of fossil fuel sponsored legislators. The problem is that we don't know how soon, if ever, reactor construction can regain the traction it once had.

Meanwhile, reactors already in operation are wearing out, breaking down, and raising public anxiety even higher. In the nuclear industry, much depends not only on venture capital but on the Atomic Energy Commission and other government agencies, and these require confidence in the product and its cost effectiveness

Since public opinion, to some extent, does stand against nuclear reactors, let me take up a few common objections briefly. Anti-nuclear movements since the 1970s have issued selected information, seldom going into IFRs or LFTRs in any detail and in some cases not at all. On the other side of the issue, Sorensen, a one-time NASA scientist and physicist at Brown Engineering, does go into nuts and bolts - as have Martin in *Superfuel* and Hargraves in *Thorium: Energy Cheaper Than Coal* (2012). They come to the same conclusion as Thomas L. Friedman does in *Hot, Flat, and Crowded* (2009), that "The threat of a nuclear leak, with today's new technology, is much less serious than the threat from climate change" (315). That doesn't address the terrorist threat that concerns the UCS, but the non-weapon-grade waste from thorium does. The repercussions of the planet reaching a tipping point make the relatively lesser danger of properly built and maintained IFRs and LFTRs a sensible choice.

More reactors will almost certainly be constructed. The questions are what kind, on what scale, at what price, and at what pace, especially in the world's largest energy consumers, Russia, China, Japan, Europe, and the US. Reactors are by no means the only answer given wind, thermal, wave, biofuel, and solar power, the latter with hopes for thermal fuel storage that captures energy from light on one side and stores it on the other where it doesn't leak out. As for natural gas (a favorite among drilling interests), its extraction releases methane, and procuring it by hydrofracking adds other serious drawbacks, including the contamination of wells and water tables. Again, public demonstrations help, as do cinema accounts of fracking like *Promised Land* (2012), focused on the community level.

Common Ground

That prisoners in the prisoner's dilemma game, under test conditions, will choose mutual benefits over immediate personal ones has been the drift of social and economic studies since Robert Axelrod's *The Evolution of Cooperation* (1984). The effects of large numbers and of teaching, however, haven't been calculated in the studies I've seen. We lack well tested concepts of cooperation at an international level, and of pedagogical efforts to broaden and strengthen it in rational ways as opposed to idea epidemics. The latter usually have proximity to carry them from person to person. National interests and foreign policy come between individual selflessness and global cooperation in encouraging people to think they are being selfless when what they are supporting is the home team, as when both sides of a senseless war praise those who participate.

That groupthink raises coordinated efforts for both good and bad causes is well established. As Nietzsche observed, behavior that is rare in individuals becomes common in crowds. Freud was aware of that too, as is Howard Bloom in *The Lucifer Principle* (1995) and *Global Brain* (2000). Casualties in the hundred-millions in the 20th century testify to it, and archaeology and history provide an abundance of older examples - as do studies of nationalist and tribal conflict like those of Winston Churchill (1931, 205), Konrad Lorenz (1966), Lawrence Keeley (1996), Keith F. Otterbein (2004), Steven LeBlanc (2003), Philip Bobbitt (2002), Donald Kagan (1995), and Azar Gat (2006). That propaganda and crowd psychology have played major roles in mass hysteria no one questions. Anthropological studies of cooperation tend to credit mainly the achievements of coordinated efforts. Films of battlefields praise the comradery and bravery and ignore the war itself as a product of that same group feeling.

A piety element among Hindu, Muslim, Jewish, Buddhist, and Christian populations often reinforces national feeling, increasing its partisan fervor and undermining the benefits of more broadly based cooperation based on what can be seen in terrain and climate. Biyearly Gallup polls since 1982 have creation and divinely-guided evolution out drawing unguided evolution about 9-1, improving in May 2012 to 32% having humans evolve under divine guidance, 46% with humans created just as they are, and 15% evolving as rock records and DNA lines of development demonstrate. The figures have improved recently, with those finding an infallible divine hand in the works outnumbering those who don't 4-1. The majority apparently isn't concerned with what that says of the Administration, given the way nature deals with its creatures, and how much of the universe is hostile to life and filled with wreckage and debris. The dodgy way, traditionally, has been to go naturalist for the bad parts and pious for the good ones – or, as we've seen, going ingenious in blaming demons, devils, and disobedient people without holding their source responsible. That versions of such a prime source vary from era to era and region to region is further reason to put them aside as a way to explain natural history, and to turn to atomic matter invariables that can be demonstrated and connected into the continuum. I'll cite just one opinion that suggests what an uphill battle it is to improve human nature. Everyone knows this, but this particular statement near the end of the last century by the polish poet Wislawa Szymborska, responding in "The Century's Decline" to idealized visions of the future, puts it simply and cogently:

> Our twentieth century was going to improve on the others.
> It will never prove it now,
> now that its years are numbered, its gait is shaky,
> its breath is short.
>
> Too many things have happened that weren't supposed to happen,
> and what was supposed to come about has not.
>
> Happiness and spring, among other things, were supposed to be getting closer.
>
> Fear was expected to leave the mountains and the valleys. Truth was supposed to hit home
> Before a lie.
>
> A couple of problems weren't going to come up anymore:
> hunger, for example, and war, and so forth.
> There was going to be respect
> for helpless people's helplessness, trust, that kind of stuff.
>
> Anyone who planned to enjoy the world is now faced
> with a hopeless task.
>
> Stupidity isn't funny.
> Wisdom isn't gay.
> Hope
> isn't that young girl anymore, et cetera, alas.

> God was finally going to believe in a man both good and strong, but good
> and strong
> are still two different men.

Except for excluding noteworthy inventions like radio, television, cars, planes, space shuttles, robots, transistors, computers – that kind of stuff – Szymborska's estimate of the 20th century seems fair enough as a correction of ill-founded optimism. A prospect that went by spans of several centuries, rather than just one, would find significant improvements over the Dark Ages: less tyranny, reduced filth and disease, and a generally higher level of education. *Strong* is now more widely distributed than, say, in the 14th century, and *good* also in the sense of the good life compared to economies based on lords and commoners, palaces and hovels. Character has a way of improving when desperation is reduced, as replacing slums with suitable housing reduces crime and disease. In matters of health, GNP, and education, the World Bank Open Data Catalogue documents a generally encouraging story, albeit seriously threatened by environmental degradation.

Szymborska's portrait of the human condition isn't cynical, merely realistic. "Since the rise of settled societies some eight to ten thousand years ago," Clive Ponting points out in *A Green History of the World* (1991), "the majority of the world's population have lived in conditions of grinding poverty. They have had few possessions, suffered from miserable living conditions and have been forced to spend most of their limited resources on obtaining enough food to stay alive" (315). Et cetera, alas, indeed. With only 10% of the world's people currently in truly dire straits, however, the educational level specifically in natural history can be raised almost universally, with the basic facts introduced at an appropriate introductory level and in greater detail at a more advanced level. Those are goals within reach, and are no more costly than current educational programs. They don't require a *re*formed human nature, only an *in*formed one. They are the equivalent of slum rejuvenation, except devoted to brainwork rather than to food and shelter, and they don't move anyone off the premises, nor do they reduce people to ill- defined souls. Brains, minds, sight, and hearing remain in operation.

The good and the strong are, after all, not the only two relevant categories, and they come in gradations. Nearly everyone has strengths and weaknesses, less good and less bad moments. The better qualities can be cultivated by the usual means of point-by-point education and include moral and aesthetic philosophy as well as natural philosophy.

Where we don't generally have to be taught affectionate attachments, reproduction, anger, anxiety, love, or fear, we do have to be taught how to distinguish salient facts from trivia and how to reach logical conclusions.

Putting critical reason to use is a learned skill; if it came automatically, fewer people would hold contradictory beliefs without discarding the impossible ones, and fewer would go into feverish fact denial as several have. Those who make realistic appraisals of what counts are less likely to be misled, less blindly chauvinistic, less bound to partisan civil, ideological, and religious orders, and less prone to neglect essential matters on behalf of lesser ones.

I mentioned cardinal points that could be posted, if not by every school district, at least by such public spirited agencies as the American Association for the Advancement of Science, the Committee on the Public Understanding of Science, Scientists' Institute for Public Information, European Initiative for Communicators of Science, and the United Nations Educational, Scientific and Cultural Organization (respectively AAAS, COPUS, SIPI, EICOS, and UNESCO) - a list I take from the commentary of Gregory and Miller on *Science in Public: Communication, Culture, and Credibility* (1998). Here are a few such points that would make my list:

- Distances extend from about 10^{-33} centimeters to billions of light years. Time extends from the briefest Planck moment (the time it takes light to travel the shortest distance, 5.39×10^{-44}s), to estimates of the future above the 10^{100}-year range.
- Astronomers, astrophysicists, and cosmologists can chronicle, with high probability, much of what has happened since the beginning of the universe. The sun and the planet collected most of their mass about 4.5 billion years ago. Life got underway about a billion years later. Since then, most of the species have gone extinct because of environments that failed them.
- According to a preponderance of scientific opinion, carbon dioxide, methane, nitrous oxide, and other greenhouse gases have increased significantly since pre industrial times. Nothing has higher priority worldwide than avoiding a runaway point in which warmer this year means still warmer next.

That isn't the kind of checklist usually presented for what the world's population needs, and in the US hasn't been prominent in those running for office until the democratic primary of 2020, when all candidates agreed that environment should have the highest priority. In 2000, the UN established millennial goals that included the eradication of extreme poverty and hunger, universal education, disease control, and worldwide cooperation on these and other matters. It included making the environment sustainable but without, as yet, an insistent concern for climate change. In considering population, resources, and sustainability from several angles, Jeffrey D. Sachs 15 years later concluded with cautious optimism that "by very careful and science-based attention to the real and growing environmental threats, we can indeed find ways to harmonize growth . . . with environmental sustainability" (182). We don't absolutely need the growth part of that, but it long ago became customary to include economic expansion in gauging national welfare. Neither it nor population growth is sustainable in the long run. Between the UN mandate in 2000 and the 2015 meeting in Paris, it became clear that climate has risen in priority in most places, even as it plunged in the US executive branch beginning in 2016. Without instruction in universal measurements and constants - combined with natural philosophy and critical reasoning – what the next 'Szymborska' report and World Bank Open Data Catalogue tell us won't likely be good news. Many of the school age population see that, but reality denial stands fast and holds high office, supported by a sizeable minority.

REFERENCES

Those familiar with the history of ideas are likely to have a healthy skepticism toward ideas that were prevalent for a time - even for a quite long time. Updates for some of what challenges the less plausible ones in physics and mathematics are posted on websites like http://arxiv.org. For current astrophysics abstracts, see *http://arxiv.org/astro-ph/new*. Several fields are represented on *sciencemag.org, scienceonline.org*, and *nature.com*. Alan Guth's appendix on units and measures (305- 309) offers scales in time, length, volume, mass, temperature, force, energy, power, frequency, and spin. The scalar spectrum for time as a whole and for what each segment measures are admirably graphic and clear in *Time in Powers of Ten: Natural Phenomena and Their Time Scale* by Gerard `t Hooft and Stefan Vandoren (2014). The *Worldwide Museum of Natural History* puts a treasure of information on two convenient charts, one condensing the history of matter and the other, a third of the total, earth history. These are available online as well as in handy highway map-size printouts. A chart of the standard model published in 2015 by Contemporary Physics Education Project (CPEP), the one reproduced in chapter nine, is also available in map size printed form and in several places on line. Variants of the standard model, time charts, and other photos of charts Googling "photos, cosmic time charts" produces, altogether some 300 pictures, diagrams, photos, and charts. These give the basic facts of the universe and, as a consequence, help us administer reality checks to notions incompatible with them.

It would be futile to try to cite all the helpful lectures and presentations on the internet, but I should mention a few on which I've drawn. In no particular order: Steven Gimble, Jeffrey C. Grossman, Michael E. Wysession, David Christian, Mark Whittle, Dan Hooper, Richard Wolfson, Steven Strogatz, Joshua Winn, Felix J. Lockman, Spencer Kelly, Sean Carroll, Don Lincoln, Lawrence Weinstein, David Ball, David Meyers, Sabine Stanley, Carlo Rovelli, and Alex Filippenko. The following references also sample hardship narratives for reminders of how muscle, brain, and nerve have coped with altitude, distances, and temperatures, in thin and in cold air, in heat, and under privation, pushing human endurance further than one would have thought possible.

Needless to say, the heat and cold that explorers have endured are but a small fraction of nature's scalar range - a good reason in itself for the lack of life in evidence anywhere in sight, except around the outer crust of our prodigiously productive satellite.

Abbey, Edward. *Desert Solitaire: A Season in the Wilderness*. New York: Simon & Schuster, 1968.

Aczel, Amir D. *God's Equation: Einstein, Relativity, and the Expanding Universe*. New York: Random House, 1999.

Adams, Frank Dawson. *The Birth and Development of the Geological Sciences*. New York: Dover, 1954.

Albritton, Claude C., Jr. *The Abyss of Time: Changing Conceptions of the Earth's Antiquity after the Sixteenth Century*. San Francisco: Freeman, Cooper & Co., 1980.

Alexander, Christopher, *et al. A Pattern Language: Towns, Buildings, Construction*. New York: Oxford University Press, 1977

Amato, Joseph A. *Dust: A History of the Small & Invisible.* Berkeley: University of California Press, 2000.

Anderson, Benedict. *Imagined Communities: Reflections on the Origin and Spread of Nationalism*. London: Verso, 1983, 2006.

Aronson, Jerrold L., Rom Harré, and Eileen Cornell Way. *Realism Rescued: How Scientific Progress Is Possible*. London: Open Court (Gerald Duckworth), 1995.

Asimov, Isaac. *Atom: Journey Across the Subatomic Cosmos*. New York: Dutton, 1991.

Auyang, Sunny Y. *How Is Quantum Field Theory Possible?* New York: Oxford University Press, 1995.

Axelrod, Robert. *The Evolution of Cooperation.* New York: Basic Books, 1984. Bacon, Francis. *A Critical Edition of the Major Works*, ed. Brian Vickers. New York: Oxford University Press, 1996.

Bacon, Francis. *Essays*, 1625. *The Proficience and Advancement of Learning, The Great Instauration, The New Organon*. Selections are conveniently gathered in *Francis Bacon: A Selection of His Works*, ed. Sidney Warhaft. New York: Macmillan, 1985.

Baggott, Jim. *Higgs: The Invention and Discovery of the 'God Particle'.* Oxford: Oxford University Press, 2012.

Baggott, Jim. *Mass: The Quest to Understand Matter from Greek Atoms to Quantum Fields*. Oxford: Oxford University Press, 2017.

Baggott, Jim. *Quantum Space: Loop Quantum Gravity and the Search for the Structure of Space, Time, and the Universe*. Oxford: Oxford University Press, 2018.

Ballahi, Mbark. *Climate Change: IPCC Report and Its Critics*. Kindle edition, 2015. Bard, Arthur S and Mitchell G. Bard. *Understanding the*

Brain. Indianapolis: Alpha Books, 2002.

Barrow, John D. *The Artful Universe*. Oxford: Oxford University Press, 1995.

Barrow, John D. *The Book of Nothing: Vacuums, Voids, and the Latest Ideas About the Origins of the Universe*. New York: Vintage Books, 2000.

Barrow, John D. *Cosmic Imagery: Key Images in the History of Science*. New York: W. W. Norton & Company, 2008.

Barrow, John D. *Theories of Everything: The Quest for Ultimate Explanation*. Oxford: Oxford University Press, 1991.

Barrow, John D. *The Universe That Discovered Itself*. Oxford: Oxford University Press, 2000.

Barrow, John D and Frank J. Tipler. *The Anthropic Cosmological Principle*. Oxford: Oxford University Press, 1986.

Bartusiak, Marcia. *The Day We Found the Universe.* New York: Random House, 2009. Bates, Marston. *The Nature of Natural History*. Princeton: Princeton University Press, 1950.

Bejan, Adrian and T. Peder Zane. *Design in Nature*. New York: Doubleday, 2012. Benuzzi, Felice. *No Picnic on Mount Kenya*. New York: E. P. Dutton, 1953.

Berreby, David. *Us and Them: Understanding Your Tribal Mind*. New York: Little, Brown and Company, 2005.

Bergreen, Laurence. *Over the Edge of the World: Magellan's Terrifying Circumnavigation of the Globe*. New York: HarperCollins, 2003.

Berry, William B. N. *Growth of a Prehistoric Time Scale*. San Francisco: W. H. Freeman, 1968.

Biddle, Wayne. *A Field Guide to the Invisible*. New York: Henry Holt & Co., 1998. Bierce, Ambrose. *The Unabridged Devil's Dictionary*, David E. Schultz and S. T. Joshi, eds. Athens, Georgia: University of Georgia Press, 2000.

Black, Jeremy and Anthony Green. *Gods, Demons and Symbols of Ancient Mesopotamia*. Austin: University of Texas Press, 1992.

Bloom, Howard. *Global Brain: The Evolution of Mass Mind from the Big Bang to the 21st Century*. New York: John Wiley & Sons, 2000.

Bloom, Howard. *The Lucifer Principle: A Scientific Expedition into the Forces of History*. New York: Atlantic Monthly Press, 1995, 1997.

Boag, Peter G. *Environment and Experience: Settlement Culture in Nineteenth-Century Oregon*. Berkeley: University of California Press, 1992.

Bobbitt, Philip. *The Shield of Achilles: War, Peace, and the Course of History*. New York: Alfred A. Knopf, 2002.

Bohr, Niels. *Atomic Physics and Human Knowledge*. Mineola, New York: Dover Publications, 2010, first pub.1961.

The Book of J. Trans. David Rosenberg, interpreted by Harold Bloom. New York: Random House, 1991.

Bowen, William A. *The Willamette Valley: Migration and Settlement on the Oregon Frontier*. Seattle: University of Washington Press, 1978.

Bowler, Peter J. *Evolution: The History of an Idea.* Berkeley: University of California Press, 2003.

Brand, Stewart. *Whole Earth Discipline: An Ecopragmatist Manifesto.* New York: Viking, 2009.

Braudel, Fernand. *On History,* trans. Sarah Matthews. Chicago: Chicago University Press, 1980.

Bregman, Rutger. *Utopia for Realists,* Elizabeth Manton, trans. New York: Little, Brown and Company, 2017.

Brenner, William H. *Wittgenstein's Philosophical Investigations.* Albany, N.Y.: State University of New York Press, 1999.

Brown, Andrew. *The Darwin Wars: How Stupid Genes Became Selfish Gods. New York: Simon and Schuster, 1999.*

Brown, Kenneth A. *Cycles of Rock and Water at the Pacific Edge.* New York: HarperCollins, 1993.

Brown, Lester R. "Could Food Shortages Bring Down Civilization?" *Lights Out*. New York: *The Scientific American, 2012.*

Brown, Lester R. *Plan B 2.0: Rescuing a Planet Under Stress and a Civilization in Trouble*. New York: W. W. Norton, 2006.

Brown, Lester R. (Earth Policy Institute). *Plan B 3.0: Mobilizing to Save Civilization*. New York: W. W. Norton.

Bryce, Robert. *Gusher of Lies: The Dangerous Delusions of 'Energy Independence'.* New York: Public Affairs, 2008.

Bryce, Robert. *Power Hungry: The Myths of 'Green' Energy and the Real Fuels of the Future*. New York: Public Affairs, 2010.

Bryce, Robert. *A Short History of Nearly Everything*. New York: Broadway Books, 2003. Buffon. *Natural History*. New York: Hurst & Co., nd.

Buhler, Brendan, "The Teeming Metropolis of You," in Dan Ariely, ed, *The Best American Science and Nature Writing*. Boston: Houghton Mifflin Harcourt, 2012.

Buller, David J. *Adapting Minds: Evolutionary Psychology and the Persistent Quest for Human Nature*. Cambridge Massachusetts: The MIT Press, 2005.

Bryce, Robert. "Four Fallacies of Pop Evolutionary Psychology," in *Becoming Human: Our Past, Present and Future*. New York: Scientific American, 2013

Burchfield, Joe D. *Lord Kelvin and the Age of the Earth*. New York: Science History, 1975.
Burke, Kenneth. *A Rhetoric of Motives*. Berkeley: University of California Press, 1950.
Burke, Kenneth. *The Rhetoric of Religion: Studies in Logology*. Berkeley: University of California Press, 1970.
Burroughs, John. *Accepting the Universe*. Kindle edition.
Burroughs, John. *The Art of Seeing Things: Essays by John Burroughs*, ed. Charlotte Walker. Syracuse, N.Y.: Syracuse University Press, 2001.
Burroughs, John. *Ways of Nature*. Boston: Houghton Mifflin, 1905.
Caras, Roger. *The Forest*. Lincoln: University of Nebraska Press, 1979.
Carroll, Sean, *The Big Picture: On the Origins of Life, Meaning, and the Universe Itself*. New York: Dutton, 2016.
Carroll, Sean. "The Higgs Boson and Beyond," 2018, "Dark Matter, Dark Energy" 2017 video lecture series.
Carroll, Sean. *Something Deeply Hidden: Quantum Worlds and the Emergence of Spacetime*. New York: Dutton, 2019.
Carusi, Annamaria and Aud Sissel. "Toward a New Ontology of Scientific Vision," *Representation in Scientific Practice Revisited*. Massachusetts Institute of Technology, 2014.
Casti, John L. *Complexification: Explaining a Paradoxical World Through the Science of Surprise*. New York: HarperCollins, 1994.
Casti, John L. *Searching for Certainty: What Scientists Can Know About the Future*. New York: Wm. Morrow, 1990.
Ceccarelli, Leah. *Shaping Science with Rhetoric*. Chicago: University of Chicago Press, 2001.
Chalmers, Alan. *The Scientist's Atom and the Philosopher's Stone: How Science Succeeded and Philosophy Failed to Gain Knowledge of Atoms*. Boston: Boston Studies in the Philosophy and History of Science, 2009.
Cherry-Garrard, Apsley. *The Worst Journey in the World*. New York: Carroll & Graf, 1989.
Chomsky, Noam. *Language and the Problems of Knowledge*. Cambridge, Mass.: The MIT Press, 1988.
Chown, Marcus. *The Magic Furnace: The Search for the Origins of Atoms*. Oxford: Oxford University Press, 2001.
Churchill, Winston S. *The World Crisis, 1911-1918*. New York: Free Press, 2005. Collier, Graham and Patricia Graham. *Antarctic Odyssey: In the Footsteps of the South Polar Explorers*. New York: Carroll & Graf, 1999.
Collier, Peter. *A Most Incomprehensible Thing: Notes Towards a Very Gentle Introduction to the Mathematics of Relativity*. Incomprehensible

books (Kindle edition), 2012.
Cone, Joseph. *Fire Under the Sea*. New York: William Morrow, 1991.
Coopmans, Catelijne, *et al*. *Representation in Scientific Practice Revisited*. Cambridge: MIT Press, 2014.
Coopmans, Catelijne, Janet Vertesi, Michael Lynch, Steve Woolgar. "Reflections" [on representation]. *Representation in Scientific Practice Revisited.* Massachusetts Institute of Technology. 2014.
Cremo, Michael A. And Richard L. Thompson. *The Hidden History of the Human Race*. Badger, California: Govardhan Hill Publishing, 1994.
Cropper, William H. *Great Physicists: The Life and Times of Leading Physicists from Galileo to Hawking*. New York: Oxford University Press, 2001.
Dalrymple, G. Brent. *The Age of the Earth*. Palo Alto: Stanford University Press, 1991. Darwin, Charles. *Charles Darwin on Evolution: The Development of the Theory of Natural Selection*, Thomas F. Glick and David Kohn, eds. Indianapolis: Hackett Publishing, 1996.
Darwin, Charles. *The Origin of the Species*. New York: Random House, 1979.
Daston, Lorraine and Katherine Park. *Wonders and the Order of Nature, 1150-1750*. New York: Zone Books, 2001.
Daston, Lorraine and Peter Galison. *Objectivity.* New York: Zone Books, 2010.
Davidson, Donald and Gilbert Harman, eds. *Semantics of Natural Language*. Boston: D. Reidel Publishing Company, 1972.
Dawkins, Richard. *The Ancestor's Tale: A Pilgrimage to the Dawn of Evolution*. New York: Houghton Mifflin, 2004.
Dawkins, Richard. *The Blind Watchmaker*. London: Penguin Books, 1988.
Dawkins, Richard. *A Devil's Chaplain: Reflections on Hope, Lies, Science, and Love.* Boston: Houghton Mifflin, 2003.
Dawkins, Richard. *The Extended Phenotype: The Long Reach of the Gene.* Oxford: Oxford University Press, 1999.
Dawkins, Richard. *The God Delusion*. New York: Houghton Mifflin, 2006.
Dawkins, Richard, ed. *The Oxford Book of Modern Science Writing*. Oxford: Oxford University Press, 2008.
Dawkins, Richard. *River Out of Eden: A Darwinian View of Life*. New York: HarperCollins, 1995.
Dawkins, Richard. *The Selfish Gene*. New York: Oxford University Press, 1976. Decartes, Rene. *The Works*. Waxkeep Publishing (Kindle edition), 2010.
Dennett, Daniel C. *Consciousness Explained.* Boston: Little, Brown and Company, 1991.

Dennett, Daniel C. *Darwin's Dangerous Idea: Evolution and the Meanings of Life.* London: Penguin, 1996.

Derrida, Jaques. *Positions*, trans. Alan Bass. Chicago: University of Chicago Press, 1981.

Derrida, Jaques. *Speech and Phenomena and Other Essays on Husserl's Theory of Signs*, trans.

David B. Allison. Evanston: Northwestern University, 1973.

Deutsch, David. *The Beginning of Infinity: Explanations that Transform the World*. New York: Penguin Books, 2011.

Deutsch, David. *The Fabric of Reality*. New York: Penguin Books, 1997.

Diamond, Jared. *Collapse: How Societies Choose to Fail Or Succeed.* New York: Viking, 2005.

Diamond, Jared. *Guns, Germs, and Steel*: *the Fates of Human Societies*. New York: W. W. Norton, 1977.

Diamond, Jared. *The Third Chimpanzee: The Evolution and Future of the Human Animal*. New York: HarperCollins, 1992.

Dirks-Edmunds, Jane Claire. *Not Just Trees: The Legacy of a Douglas-fir Forest.* Pullman, Washington: Washington State University Press, 1999.

Donald, Merlin. *A Mind So Rare*: *The Evolution of Human Consciousness.* New York: W. W. Norton, 2001.

Donald, Merlin. *Origins of the Modern Mind.* Cambridge: Harvard University Press, 1991. Dougherty, Jude P. *The Nature of Scientific Explanation*. Washington, D.C.: The Catholic University of America Press, 2013.

Dunbar, Robin. *The Trouble with Science*. Cambridge, Mass.: Harvard University Press, 1995.

Dunham, William. *Journey through Genius: the Great Theorems of Mathematics*. New York: Penguin, 1991.

Durant, Will & Ariel. *The Lessons of History*. New York: Simon & Schuster, 1968. Dyson, Freeman. *Infinite in All Directions*. New York: Harper & Row, 1989.

Edelman, Gerald M. *Bright Air, Brilliant Fire: On the Matter of the Mind.* New York: Basic Books, 1992.

Edelman, Gerald M. *Wider Than the Sky: The Phenomenal Gift of Consciousness*. New Haven: Yale University Press, 2004.

Edmonds, David and John Eidinow. *Wittgenstein's Poker*. New York: Ecco, 2002. Ehrlich, Paul R. *Human Nature: Genes, Cultures, and the Human Prospect*. New York: Penguin Books, 2002.

Einstein, [Albert] *et al. The Principle of Relativity*. New York: Dover Publications, 1952.

Einstein, [Albert]. *Relativity: The Special and the General Theory*. New Delhi: General Press, 2012. Eldredge, Niles. *The Monkey Business: A Scientist Looks at Creationism*. New York: Washington Square Press, 1982.

Eliade, Mircea. *The Myth of the Eternal Return* (1954), Willard R. Trask, trans. Princeton: Princeton University Press, 2005.

Eliade, Mircea. *The Sacred and the Profane: The Nature of Religion* (1957), Willard R. Trask, trans. New York: Harcourt Brace & Company.

Ellis, John M. *Against Deconstruction*. Princeton: Princeton University Press, 1989. Emspak, Jesse. "The Four Horsmen," *Lights Out*. New York: The Scientific American, 2012.

EPA: Office of Atmospheric Programs. *Methane and Nitrous Oxide Emissions From Natural Sources.* Washington, D.C., 2010. Online in www.epa.gov/methane/- pdfs/Methane-and-Nitrous-Oxide-Emissions-From-Natural-Sources.pdf

Epicurus. *The Essential Epicurus: Letters, Principal Doctrines, Vatican Sayings, and Fragments*, Eugene O'Connor, trans. Amherst, New York: Prometheus Books, 1993.

Fairclough, Norman. *Discourse and Social Change*. Cambridge: Polity Press, 1992. Ferris, Timothy. *Coming of Age in the Milky Way*. New York: HarperCollins Publishers, 2003.

Ferris, Timothy. *The Mind's Sky: Human Intelligence in a Cosmic Context*. New York: Bantam Books, 1992.

Feyerabend, Paul. *Farewell to Reason*. London: Verso, 1987.

Feyerabend, Paul. *Realism, Rationalism and Scientific Method*, Philosophical Papers, vol. 1. Cambridge: Cambridge University Press, 1981.

Feynman, Richard. *The Character of Physical Laws*. New York: The Modern Library, 1994.

Feynman, Richard. *The Meaning of It All*. Reading, Massachusetts: Perseus Books, 1998.

Feynman, Richard. *QED: The Strange Theory of Light and Matter*. Princeton: Princeton University Press, 1985.

Feynman, Richard. *Six Not-So-Easy Pieces: Einstein's Relativity, Symmetry, and Space-Time*. Cambridge: Perseus Books, 1997.

Fine, Arthur, "The Einstein-Podolsky-Rosen Argument in Quantum Theory", *The Stanford Encyclopedia of Philosophy* (Winter 2017 Edition), Edward N. Zalta (ed.), URL = <https://plato.stanford.edu/archives/win2017/entries/qt-epr/>

Fisch, Harold. *Poetry With Purpose: Biblical Poetics and Interpretation*. Bloomington: Indiana University Press, 1990.

Flannery, Timothy. *An Ecological History of North America and Its Peoples*. London: William Heinemann, 2001.

Forster, Piers and Venkatachalam Ramaswamy, "Changes in Atmospheric Constituents and in Radiative Forcing," chapter two of the IPCC fifth assessment report on climate change.

Foucault, Michel. *The Archaeology of Knowledge*, trans. A. M. Sheridan Smith. New York: Random House, 1972

Foucault, Michel. *The Order of Things: An Archaeology of the Human Sciences*. New York: Random House, 1970.

Fortey, Richard. *Life: An Unauthorised Biography*. London: HarperCollins, 1998. Friedman, Thomas L. *Hot, Flat, and Crowded: Why We Need a Green Revolution*. New York: Picador / Farrar, Straus and Giroux, 2009.

Galileo Galilei. *The Essential Galileo*, Maurice A. Finocchiaro, ed. and trans. Indianapolis: Hackett publishing Company, Inc., 2008.

Gamow, George. *Gravity.* Mineola, N.Y.: Dover Publications, 2002.

Gamow, George. *Thirty Years That Shook Physics: The Story of Quantum Theory.* New York: Dover, 1966.

Gamow, George. *Mr. Tompkins in Paperback.* Cambridge: Cambridge University Press, 1965. Gardner, Martin. *Fads and Fallacies*. New York: Dover, 1957.

Gamow, George. *The Night is Large*. New York: St. Martin's Griffin, 1996.

Gamow, George. *The Whys of a Philosophical Scrivener*. New York: St. Martin's Griffin, 1999. Gat, Azar. *War in Human Civilization*. Oxford: Oxford University Press, 2006.

Gilbar, Steven ed. *Natural State: A Literary Anthology of California Nature Writing*. Berkeley: University of California Press, 1998.

Goldberg, Dave. *The Universe in the Rearview Mirror: How Hidden Symmetries Shape Reality*. New York: Penguin Group, 2013.

Goldberg, Dave and Jeff Blomquist. *A User's Guide to the Universe: Surviving the Perils of Black Holes, Time Paradoxes, and Quantum Uncertainty*. Hoboken, N.J.: John Wiley & Sons, 2010.

Good, Gregory A., ed. *Sciences of the Earth: An Encyclopedia of Events, People, and Phenomena*. New York: Garland Publishing, 1998.

Gore, Al. *Earth in the Balance: Ecology and the Human Spirit*. New York: Penguin, 1992.

Gould, Stephen Jay. *Dinosaur in a Haystack.* New York: Crown Trade, 1996.

Gould, Stephen Jay. *Eight Little Piggies.* New York: W. W. Norton, 1993.

Gould, Stephen Jay. *Ever Since Darwin*. New York: W. W. Norton, 1977.

Gould, Stephen Jay. *Full House: The Spread of Excellence from Plato to Darwin*. New York: Three Rivers Press, 1996.

Gould, Stephen Jay. *The Lying Stones of Marrakech: Penultimate Reflections in Natural History*. New York: Harmony Books, 2000.
Gould, Stephen Jay. *The Structure of Evolutionary Theory*. Cambridge, Massachusetts: Harvard University Press, 2002.
Gould, Stephen Jay. *Wonderful Life: The Burgess Shale and the Nature of History*. New York: W. W. Norton, 1989.
Greenblatt, Stephen. *The Swerve: How the World Became Modern*. New York: W. W. Norton & Company, 2011.
Greene, Brian. *The Elegant Universe: Superstrings, Hidden Dimensions, and the Quest for the Ultimate Theory*. New York: Random House, 1999.
Greene, Brian. *The Fabric of the Universe*: *Space, Time, and the Fabric of the Universe*. New York: Random House, 2007.
Greene, Brian. *The Hidden Reality: Parallel Universes and the Deep Laws of the Cosmos*. New York: Random House, 2011.
Green, Michael. *Quantum Physics and Ultimate Reality: Mystical Writings of Great Physicists*. Kindle, 2013.
Greer, John Michael. *Not the Future We Ordered: Peak Oil, Psychology, and the Myth of Progress*. London: Karnac Books, 2013.
Gregory, Jane and Steve Miller. *Science in Public: Communication, Culture, and Credibility*. Cambridge, Mass.: Perseus, 1998.
Gribbin, John. *Almost Everyone's Guide to Science*. New Haven: Yale University Press, 1999.
Gribbin, John and Michael White. *Stephen Hawking: A Life in Science*. New York: Penguin, 1993.
Gribbin, John and Mary Gribbin. *Fire on Earth: Doomsday, Dinosaurs, and Humankind.* New York: St. Martin's Press, 1996.
Gross, Alan G. *The Rhetoric of Science*. Cambridge, Mass.: Harvard University Press, 1990.
Gross, Alan G_and William M. Keith, eds. *Rhetorical Hermeneutics: Invention and Interpretation in the Age of Science*. Albany: State University of New York Press, 1997.
Gross, Alan G and Joseph E. Harmon. *Science from Sight To Insight: How Scientists Illustrate Meaning.* Chicago: University of Chicago Press, 2014.
Guterl, Fred. *The Fate of the Species.* New York: Bloomsbury, 2012.
Guth, Alan H. *The Inflationary Universe*. Reading, Massachusetts: Helix Books. 1997. Haber, Francis C. *The Age of the World: Moses to Darwin*. Baltimore: Johns Hopkins University Press, 1959.
Haken, H., ed. *Chaos and Order in Nature*. Heidelberg: Springer-Verlag Berlin, 1981. Hall, Nina, ed. *Exploring Chaos: A Guide to the New*

Science of Disorder. New York: Norton, 1991.
Hargraves, Robert. *Thorium: Energy Cheaper Than Coal.* Kindle, 2012. Harris, Sam. *Letter To a Christian Nation*. New York: Alfred Knopf, 2006.
Harris, Sam. *The End of Faith: Religion, Terror, and the Future of Reason.* New York: ton, 2005.
Harris, Stephen L. *Agents of Chaos*. Missoula: Mountain Press, 1990.
Harris, Stephen L. *Fire Mountains of the West*. Missoula, Montana: Mountain Press, 1988. Harte, John. *The Green Fuse: An Ecological Odyssey*. Berkeley: University of California Press, 1993.
Hartmann, Thom. *Last Hours of Humanity: Warming the World to Extinction*. Kindle edition, 2013.
Hartmann, William K. And Ron Miller. *The History of Earth: An Illustrated Chronicle of an Evolving Planet*. New York: Workman Publishing, 1991.
Hawken, Paul, Amory Lovins and L. Hunter Lovins, *Natural Capitalism: Creating the Next Industrial Revolution.* Boston: Little, Brown and Company.
Hawking, Stephen. *Black Holes and Baby Universes*. New York: Bantam Books, 1993.
Hawking, Stephen. *A Brief History of Time*. New York: Bantam Books, 1996.
Hawking, Stephen. *The Universe in a Nutshell*. New York: Bantom Books, 2001.
Hayek, F. A. *The Road to Serfdom*, vol. 2, Bruce Caldwell, ed. Chicago: University of Chicago Press.
Hilbert, David. *The Fundations of Geometry*, trans. E. J. Townsend. Read Books Ld., 2011.
Hogan, J. Michael, ed. *Rhetoric and Community: Studies in Unity and Fragmentation*. Columbia, S.C.: University of South Carolina Press, 1998.
Heim, Michael. *The Metaphysics of Virtual Reality*. New York: Oxford University Press, 1993.
Heinberg, Richard. *The Party's Over: Oil, War and the Fate of Industrial Societies*. Gabriola Island, BC: New Society Publishers, 2003.
Heinberg, Richard. *Snake Oil: How Fracking's False Promise of Plenty Imperils Our Future*. Santa Rosa, California: Post Carbon Institute, 2013.
Holbach, Paul Henri Thiry. *The System of Nature*, part II, Samuel Wilkinson, trans. Public domain book.

Holmes, Arthur. *Principles of Physical Geology*, 3rd ed., revised by Doris L. Holmes. New York: John Wiley & Sons, 1978.

Holmgren, David. *Future Scenarios: How Communities Can Adapt to Peak Oil and Climate Change*. White River Junction, VT: Chelsea Green Publishing Co., 2009.

Hooft, Gerard 't and Stefan Vandoren. *Time in Powers of Ten: Natural Phenomena and Their Timescales*. London: World Scientific Publishing, 2014.

Hooke, Robert. *Micrographia: Or Some Physiological Descriptions of Minute Bodies Made by Magnifying Glasses with Observations and Inquiries Thereupon.* London: Jo. Martyn and J. Allestry, 1665.

Hooper, Dan, "What Einstein Got Wrong" (2017), video lecture series.

Horwitz, Tony. *Blue Latitudes: Boldly Going Where Captain Cook Has Gone Before*. New York: Henry Holt, 2002.

Howarth, Robert W. *et al. Venting and leaking of methane from shale gas development: response to Cathles et al. 2011.*

Hutton, James. *Theory of the Earth* (1788). New York: Classic Books International, 2010,

Illingworth, Valerie, ed. *The Penguin Dictionary of Physics*. London: Penguin, 1990. IPCC, 2013: Summary for Policymakers. In: *Climate Change 2013: The Physical Science Basis. Contribution of Working Group I to the Fifth Assessment Report of the Intergovernmental Panel on Climate Change.*

IPCC website, http://www.ipcc.ch/pdf/assessment-report/ar4/wg1/ar4-wg1-chapter2.pdf concerning "Changes in Atmospheric Constituents in Radiative Forcing."

Irwin, Alan and Brian Wynne, eds. *Misunderstanding Science? The Public Reconstruction of Science and Technology*. Cambridge: Cambridge University Press, 1996.

Jackendoff, Ray. *Foundations of Language: Brain, Meaning, Grammar, Evolution*. Oxford: Oxford University Press, 2002.

James, William. *Essays in Radical Empiricism and A Pluralistic Universe*. Glouchester, Mass.: Peter Smith, 1967.

James, William. James, William. *Pragmatism and Four Essays from the Meaning of Truth*. Cleveland and New York: Meridian Books, 1955.

Jeans, James. *The Mysterious Universe*. Cambridge: Cambridge University Press, 1930. Jayardwardhana, Ray. *Neutrino Hunters*. New York: Farrar, Straus and Giroux, 2013.

Johanson, Donald, and Blake Edgar. *From Lucy to Language*. New York: Simon & Schuster, 1996.

Jones, Steve. *Almost Like a Whale: The Origin of Species Updated*. London: Doubleday, 1999.

Kagan, Donald. *On the Origins of War: and the Preservation of Peace*. New York: Doubleday, 1995.

Kagan, Donald. *The Outbreak of the Peloponnesian War*. Ithaca: Cornell University Press, 1969.

Kahn, Paul W. *Sacred Violence, Torture, Terror, and Sovereignty*. Ann Arbor: University of Michigan Press, 2008.

Kaku, Michio. *Hyperspace: A Scientific Odyssey through Parallel Universes, Time Warps, and the 10th Dimension*. New York: Oxford University Press, 1994.

Kaku, Michio. *Physics of the Future: How Science Will Shape Human Destiny and Our daily Lives by the Year 2100.* New York: Doubleday, 2011.

Kane, Gordon. *Supersymmetry: Unveiling the Ultimate Laws of Nature*. Cambridge, Massachusetts: Perseus Publishing, 2000.

Kant, Immanuel. *The Critique of Pure Reason*, J.M.D. Meiklejohn, trans. Amazon digital services, 1787 edition.

Kaufmann, Eric. *Shall the Religious Inherit the Earth? Demography and Politics in the Twenty-first Century.* London: Profile Books Ltd. 2010,

Kaufman, Maynard. *Adapting to the End of Oil*. Xlibris, 2008.

Keeley, Lawrence H. *War Before Civilization: The Myth of the Peaceful Savage*. New York: Oxford University Press, 1996.

Kirshner, Robert P. *The Extravagant Universe: Exploding Stars, Dark Energy, and the Accelerating Cosmos.* Princeton: Prinreton University Press, 2002.

Klinkenborg, Verlyn. *The Rural Life.* Boston: Little, Brown and Company, 2003. Kolbert, Elizabeth. *The Sixth Extinction: An Unnatural History.* New York: Henry Holt and Company, 2014.

Korten, David C. *The Post-Corporate World: Life After Capitalism*. West Hartford, CT: Kumarian Press, 1999.

Krakauer, Jon. *Into Thin Air*. New York: Anchor, 1998.

Kress, Gunther. *Multimodality: A Social Semiotic Approach to Contemporary Communication.* New York: Routledge, 2010.

Krieger, Murray, ed. *The Aims of Representation*. Stanford: Stanford University Press, 1993.

Kroll, Richard, Richard Ashcraft, and Perez Zagorin. *Philosophy, Science, and Religion in England, 1640-1700*. Cambridge: Cambridge University Press, 1992.

Kuhn, Thomas. *The Structure of Scientific Revolutions*. Chicago: University of Chicago Press, 1970.

Kumar, Manjit. *Quantum: Einstein, Bohr, and the Great Debate About the Nature of Reality*. New York: W. W. Norton, 2008, 2011.

Kunstler, James Howard. *The Long Emergency: Surviving the End of Oil, Climate Change, and Other Converging Catastrophes of the Twenty-First Century*. New York: Grove Press, 2006.

Kunstler, James Howard. *Too Much Magic: Wishful Thinking, Technology, and the Fate of the Nation.* New York: Atlantic Monthly Press, 2012.

Lakoff, George. "Linguistics and Natural Logic," in *Semantics of Natural Language*, Donald Davidson and Gilbert Hartman, eds. Boston: D. Reidel Publishing Company, 1972.

Lakoff, George and Mark Johnson. *Metaphors We Live By*. Chicago: University of Chicago Press, 2003.

Lakoff, George. *Women, Fire, and Dangerous Things: What Categories Reveal about the Mind*. Chicago: University of Chicago Press, 1987.

Lanham, Richard. *The Electronic Word: Democracy, Technology, and the Arts*. Chicago: University of Chicago Press, 1993.

Leakey, Richard. *The Origin of Humankind*. New York: HarperCollins, 1994. LeBlanc, Steven with Katherine E. Register. *Constant Battle: The Myth of the Peaceful, Noble Savage.* New York: St. Martin's Press, 2003.

Le Bon, Gustave. *The Crowd: Study of the Popular Mind*. San Bernadino, California, 2013.

Lederman, Leon and Christopher Hill. *Beyond the God Particle*. Amherst, New York: Prometheus Books, 2013.

Lederman, Leon and Christopher Hill. *The God Particle*. New York: Bantam Doubleday, 1994.

Lederman, Leon and Christopher Hill and Christopher T. Hill. *Symmetry and the Beautiful Universe*. Amherst, New York: Prometheus Books, 2004.

Leopold, Aldo. *A Sand County Almanac*. New York: Oxford University Press,1949. Lewis, Cherry. *The Dating Game: One Man's Search for the Age of the Earth*. Cambridge: Cambridge University Press, 2000.

Lightman, Alan. *The Discoveries*. New York: Pantheon, 2005.

Lightman, Alan. *A Sense of the Mysterious: Science and the Human Spirit*. New York: Random House, 2005.

Lincoln, Don, "The Theory of Everything: The Quest to Explain All Reality," 2018 video lecture course.

Lloyd, G.E.R. *Early Greek Science: Thales to Aristotle*. New York: W. W. Norton and Company, 1970.

Lloyd, G.E.R. Greek *Science After Aristotle*. New York: W. W. Norton and Company, 1973. Lorentz, H.A., A. Einstein, H. Minkowsky, H. Weyl.

The Principle of Relativity. W. Perrett and G. B. Jeffery, trans. New York: Dover Publications, 1923, 1952.

Lorenz, Konrad. *On Aggression*, Marjorie Kerr Wilson, trans. New York: Bantam, 1966.

Losee, John. *A Historical Introduction to the Philosophy of Science.* Oxford: Oxford University Press, 2001.

Lucretius. *The Way Things Are, The* de rerum natura *of Titus Lucretius Carus*, Rolfe Humphries, trans. Bloomington: Indiana University press, 1968.

Lyell, Charles. *Principles of Geology*. London: Penguin Books, 1997 (1830-1833). Lynas, Mark. *Nuclear 2.0: Why a Green Future Needs Nuclear Power*. Kindle Single, 2013.

Lyon, Thomas J, ed. *This Incomperable Lande: A Book of American Nature Writing*. New York: Penguin Books, 1989.

Mackay, Charles. *Extraordinary Popular Delusions and the Madness of Crowds.* London: Richard Bentley, 1841, and Seven Treasures Publications, 2008.

McDougall, Walter A. *The Heavens and the Earth: A Political History of the Space Age*. Baltimore: Johns Hopkins University Press, 1997 (first published 1985).

McGhee, George R., Jr. *The Late Devonian Mass Extinction: The Frasnian/Famennian Crisis*. New York: Columbia University Press, 1996.

McGuire, Bill. *A Guide to the End of the World: Everything You Never Wanted to Know*. Oxford: Oxford University Press, 2002.

McKibben, Bill. *Eaarth: Making a Life on a Tough New Planet*. New York: Henry Holt, 2010.

McKinney, H. Lewis, ed. *Lamarck to Darwin: Contributions to Evolutionary Biology.* Lawrence, Kansas: Coronado Press, 1971.

McPhee, John. *Annals of the Former World.* New York: Farrar, Straus and Giroux, 1998.

McPhee, John. *Assembling California*. New York: Noonday Press, 1993. Mailloux, Steven. *Rhetorical Power.* Ithaca: Cornell University Press, 1989.

Mangalwadi, Vishal. *The Book That Made Your World: How the Bible Created the Soul of Western Civilization*. Nashville: Thomas Nelson, 2011.

Martin, Richard. *Super Fuel: Thorium, the Green Energy Source for the Future*. New York: St. Martin's Press, 2012.

Maser, Chris. *Forest Primeval: The Natural History of an Ancient Forest.* Corvallis, Oregon: Oregon State University Press, 1989

Mayr, Ernst. *Systematics and the Origin of Species*. New York: Columbia University Press, 1942, 1999.

Medawar, Peter. *Pluto's Republic*. New York: Oxford University Press, 1984. Meyers, Carol. *Rediscovering Eve: Ancient Israelite Women in Context*. New York: Oxford University Press, 2013.

Myers, Greg. "Discourse Studies of Scientific Popularization: Questioning the Boundaries," *Discourse Studies* 5 (2003), 265-279.

Millard, Candice. *The River of Doubt: Theodore Roosevelt's Darkest Journey*. New York: Anchor Books, 2006.

Mills, David. *Atheist Universe: The Thinking Persons Answer to Christian Fundamentalism*. Berkeley: Ulysses Press, 2006.

Milton, John. *Complete Poems and Major Prose*, ed. Merritt Y. Hughes. New York: Macmillan, 1957.

Milosz, Czeslaw. *Road-Side Dog*. New York: Farrar, Straus and Giroux, 1998. Montague, Ashley, ed., *Man and Aggression*. London: Oxford University Press, 1973. Mitchell, Melania. *Complexity: A Guided Tour*. Kindle edition, 2009.

Mithen, Steven. *The Prehistory of the Mind*. London: Thames and Hudson, 1996. Monbiot, George. *Heat: How To Stop the Planet from Burning*. Cambridge, Mass.: South End Press, 2007.

Morano, Mark. *Climate Depot*. Climatedepot.com. 2013.

Morris, Ian. *Why the West Rules–for Now*. New York: Farrar, Straus and Giroux, 2010. Morris, Simon Conway. *The Crucible of Creation: The Burgess Shale and the Rise of Animals.* Oxford: Oxford University Press, 1998.

Muir, John. *The Mountains of California*. New York: Dorset Press, 1988 (originally 1894)

Murray, John A., ed. *American Nature Writing, 1995*. San Francisco: Sierra Club Books, 1995.

Musser, George, "Could Time End" in *Lights Out: How It All Ends* and in *A Question of Time: The Ultimate Paradox*. New York: Scientific American, 2012.

Nagel, Ernest and James R. Newman. *Gödel's Proof.* New York: New York University Press, 1958.

Nerlich, Steve. *Astronomy Without a Telescope*. Kindle Edition, 2012.

Newton, Isaac. *The Principia: Mathematical Principles of Natural Philosophy*, I. Bernard Cohen and Anne Whitman, trans. Berkeley: University of California Press, 1999.

Newton-Smith, W. H. *The Rationality of Science*. London: Routledge, 1981.

Nicolson, Iain. *Dark Side of the Universe: Dark Matter, Dark Energy, and the Fate of the Cosmos*. Baltimore: Johns Hopkins University Press,

2007.
Nicolson, Margorie. *Mountain Gloom and Mountain Glory*. Seattle: University of Washington Press, 1997.
Nordhaus, Ted and Michael Shellenberger. *Break Through: Why We Can't Leave Saving the Planet to Environmentalists*. New York: Houghton Mifflin Harcourt, 2009.
Norenzayan, Ara. *Big Gods: How Religion Transformed Cooperation and Contention*. Princeton: Princeton University Press, 2013.
Nowak, Martin A. With Roger Highfield. *Super Cooperator: Altruism, Evolution, and WhyWeNeed Each Other to Succeed.* New York: Free Press, 2011.
Oldroyd, David. *Thinking About the Earth: A History of Ideas in Geology*. London: Athlone, 1996.
Omerod, Paul. *Why Most Things Fail and How to Avoid It*. London: Faber and Faber, 2005.
Ong, Walter J. *Interfaces of the Word: Studies in the Evolution of Consciousness and Culture*. Ithaca: Cornell University Press, 1977.
Ophuls, William. *Immoderate Greatness: Why Civilizations Fail*. North Charleston, SC, 2012.
Oreskes, Naomi. *The Rejection of Continental Drift: Theory and Method in American Earth Science.* New York: Oxford University Press, 1999.
Oerter, Robert. *Theory of Almost Everything. The Standard Model. The Unsung Triumph of Nuclear Physics.* New York: Pearson Education, 2006.
Osserman, Robert. *Poetry of the Universe: A Mathematicall Exploration of the Cosmos*. New York: Random House, 1995.
Otterbein, Keith F. *How War Began*. College Station, Texas: Texas A & M Press, 2004. Pasachoff, Jay M. and Alex Filippenko. *The Cosmos: Astronomy in the New Millennium*. Pacific Grove, CA.: Brooks/Cole division of Thomson, 2004 Pauwels, Luc, ed. *Visual Cultures of Science: Rethinking Representation Practices in*
Knowledge Building and Science Communication. Hanover and London: University Press of New England, 2006.
Peattie, Donald Culross. *An Almanac for Moderns*. New York: G. P. Putnam's Sons, 1935.
Pérez-Llantada, Carmen. *Scientific Discourse and the Rhetoric of Globalization: The Impact of Culture and Language.* London: Continuum International Publishing Group, 2012.
Phillips, Kevin. *The Politics of Rich and Poor*. New York: Harper, 1991.
Pinker, Steven. *How the Mind Works*. New York: W. W. Norton & Company, 1997.

Pinker, Steven. *The Stuff of Thought: Language As a Window into Human Nature*. London: Penguin, 2007.

Plait, Philip. *Death from the Skies: These Are the Ways the World Will End*. New York: Penguin Group, 2008.

Playfair, William. *An Inquiry into the Permanent Causes of the Decline and Fall of Powerful and Wealthy Nations*. London: Greenland and Norris, 1807.

Pollack, Robert. *The Missing Moment: How the Unconscious Shapes Modern Science*. New York: Houghton Mifflin, 1999.

Pollard, Justin and Howard Reid. *The Rise and Fall of Alexandria, Birthplace of the Modern Mind*. Tanto Media, 2006.

Ponting, Clive. *A Green History of the World: The Environment and the Collapse of Great Civilizations*. New York: Penguin Books, 1991.

Popper, Karl. *The Logic of Scientific Discovery*. London: Routledge, 1959, 2002. Powell, James Lawrence. *Mysteries of Terra Firma: The Age and Evolution of the Earth*. New York: Simon & Schuster, 2001.

Powell, James Lawrence. *Night Comes to the Cretaceous*. New York: Harcourt Brace, 1998.

Pringle, Heather. *In Search of Ancient North America: An Archaeological Journey to Forgotten Cultures*. New York: John Wiley & Sons, 1996.

Putnam, Hilary. *Representation and Reality*. Cambridge, Mass.: MIT Press, 1988.

Putnam, Hilary. *The Threefold Cord: Mind, body, and World*. New York: Columbia University Press, 1999.

Quine, W. V. *The Pursuit of Truth*. Cambridge, Mass.: Harvard University Press, 1992.

Raff, Rudolf A. *The Shape of Life: Genes, Development, and the Evolution of Animal Form*. Chicago: University of Chicago Press, 1996.

Ratcliffe, Martin. *State of the Universe 2008*. Chichester, UK: Praxis Publishing Ltd., 2008.

Rawicz, Slavomir. *The Long Walk*. Lyons Press, 1997.

Redhead, Michael. "Should We Believe in Quarks and QCD?" (637-644), in *The Rise of the Standard Model: Particle Physics in the 1960s and 1970s*, Lillian Hoddeson, et al, eds. Cambridge: Cambridge University Press, 1997.

Redman, Charles. *The Rise of Civilization: From Early Farmers to Urban Society in the Ancient Near East*. San Francisco: W. H. Freeman, 1978.

Rees, Martin. *Just Six Numbers: The Deep Forces That Shape the Universe*. London: Phoenix, 2000.

Regis, Ed. *Who Got Einstein's Office?* Reading, Massachusetts: Perseus Books, 1987. Riordan, Michael. *The Hunting of the Quark: A True Story*

of Modern Physics. New York: Simon & Schuster, 1987.
Rocky Mountain Institute. *Annual Report: Speeding the Energy Revolution*. Carbon War Room, 2014-2015.
Roessler, Johannes, Hemdat Lerman, and Naomi Eilan, eds. *Perception, Causation, and Objectivity*. New York: Oxford University Press, 2011.
Rössler, O. E., "How Chaotic Is the Universe?" in Arun V. Holden, ed., *Chaos*. Princeton: Princeton University Press, 1986.
Rucker, Rudy. *Infinity and the Mind: The Science and Philosophy of the Infinite*. Princeton: Princeton University Press, 1995.
Ruddiman, William F. *Plows, Plagues, and Petroleum: How Humans Took control of Climate*. Princeton: Princeton University Press, 2010.
Ruse, Michael. *Darwin and Design: Does Evolution Have a Purpose?* Cambridge, Massachusetts: Harvard University Press, 2003.
Russell, Bertrand. *The Problems of Philosophy*. New York: Galaxy Book, 1959.
Russell, Bertrand. *Religion and Science*. New York: Oxford University Press, 1997. First published, 1935.
Rykwert, Joseph. *The Seduction of Place: The History and Future of the City*. New York: Random House, 2002.
Ryle, Gilbert. *The Concept of Mind*. Chicago: Chicago University Press, 1984 (originally 1949).
Sachs, Jeffrey D. *The Age of Sustainable Development*. New York: Columbia University Press, 2015.
Sagan, Carl. *Cosmos*. New York: Wings Books, 1980.
Sagan, Carl. *The Demon-Haunted World: Science as a Candle in the Dark*. New York: Ballantine Books, 1996.
Sagan, Carl. *Pale Blue Dot*. New York: Random House, 1994.
Sample, Ian. *Massive: The Missing Particle That Sparked the Greatest Hunt in Science*. New York: Basic Books, 2010.
Saner, Reg. *The Four-Cornered Falcon: Essays on the Interior West and the Natural Scene*. Baltimore: The Johns Hopkins University Press, 1993.
Sarles, Harvey B. *Language and Human Nature: Toward a Grammar of Interaction and Discourse*. Minneapolis: University of Minnesota Press, 1985.
Sartori, Leo. *Understanding Relativity: A Simplified Approach to Einstein's Theories*. Berkeley: University of California Press, 1996.
Schama, Simon. *Landscape and Memory*. New York: Alfred Knopf, 1995.
Schatzman, Evry. *Our Expanding Universe* New York: McGraw-Hill, 1992. Schoenherr, Allan A. *A Natural History of California*. Berkeley: University of California Press, 1992.

Schroeder, Karl. *Permanence.* New York: Tom Doherty Associates, 2002.
Schweber, Silvan, "Historical Perspective on the Rise of the Standard Model," in Lillian Hoddelson, *et al,* eds., *The Rise of the Standard Model: Particle Physics in the 1960s and 1970s*. Cambridge: Cambridge University Press, 1997.
Schumm, Bruce A. *Deep Down Things: The Breathtaking Beauty of Particle Physics*. Baltimore: Johns Hopkins University Press, 2004.
Schwantes, Carlos A. *The Pacific Northwest: An Interpretive History*. Lincoln: University of Nebraska Press, 1989.
Schwartz, Jeffrey H. *Sudden Origins: Fossils, Genes, and the Emergence of Species*. New York: John Wiley and Sons, 1999.
Schwartz, Joseph. *The Creative moment: How Science Made Itself Alien to Modern Culture*. New York: HarperCollins, 1992.
Schweber, Silvan, "A Historical Perspective on the Rise of the Standard Model," in Lillian Hoddeson, et al, eds., *The Rise of the Standard Model: Particle Physics in the 1960s and 1970s.* Cambridge: Cambridge University Press, 1997.
Scientific American, editors. *Seeking the Multiverse*. New York: The Scientific American, 2017.
Scientific American, editors. *Lights Out: How It All Ends*. New York: The Scientific American, 2012.
Searle, John R. *Mind, Language, and Society: Philosophy in the Real World.* New York: Basic Books, 1998.
Scientific American, Remember When: The Science of Memory. New York: 2013.
Searle, John R. *The Secrets of Consciousness.* Kindle edition.
Shackleton, Ernest. *The Heart of the Antarctic*. New York: Carroll & Graf, 1999.
Shackleton, Ernest. *South*. New York: Carroll & Graf, 1998.
Shankar, R. *Fundamentals of Physics: Mechanics, Relativity, and Thermodynamics.* New Haven: Yale University Press, 2014.
Shore, Willmam H., ed. *The Nature of Nature*. New York: Harcourt Brace & Company, 1994.
Sides, Hampton. *In the Kingdom of Ice: The Grand and Terrible Polar Voyage of the* USS Jeannette. New York: Doubhleday, 2014.
Simon, Alvah. *North to the Night: A Year in the Arctic Ice*. New York: McGraw Hill, 1999.
Simons, Herbert W., ed. *Invention and Persuasion in the Conduct of Inquiry*. Chicago: University of Chicago Press, 1990.
Simpson, Joe. *Touching the Void.* London: Jonathan Cape, 1988. Singh, Simon. *Fermat's Enigma*. New York: Random House, 1997.

Smith, Cameron M. and Charles Sullivan. *The Top 10 Myths About Evolution*. Prometheus Books: New York, 2007.

Smith, Gar. *Nuclear Roulette: The Truth About the Most Dangerous Energy Source on Earth*. White River Junction, VT: Chelsea Green Publishing, 2012.

Smith, Nicholas and Helene Joffe. "How the public Engages with Global Warming: A Social Representations Approach," *Public Understanding of Science* 22 (2012), 16- 32.

Smolin, Lee. *Three Roads to Quantum Gravity*. New York: Basic Books, 2001.

Snow, C. P. *The Two Cultures*. Cambridge: Cambridge University Press, 1998. First published in 1959.

Sokal, Alan and Jean Bricmont. *Fashionable Nonsense: Postmodern Intellectuals' Abuse of Science*. New York: Picador USA, 1998.

Starke, Linda, ed. *The State of the World*. New York: Norton/Worldwatch, 2006. Stegner, Wallace. *Wolf Willow: A History, A Story, and a Memoir of the Last Plains Frontier*. New York: Penguin, 1990. First published, 1962.

Steinhardt, Paul J. and Neil Turok. *Endless Universe: Beyond the Big Bang*. New York: Doubleday, 2007.

Stephen, Sir Leslie, and Sir Sidney Lee, eds. *Dictionary of National Biography*. Oxford: Oxford University Press, 1973.

Stevens, Wallace. *Letters of Wallace Stevens*, Holly Stevens, ed. Berkeley: University of California Press.

Stevens, Wallace. *The Necessary Angel: Essays on Reality and the Imagination*. New York: Random House, 1942.

Stevens, Wallace. *Opus Posthumous*. New York: Random House, 1957.

Stevens, Wallace. *The Palm At the End of the Mind*, ed. Holly Stevens. New York: Alfred A. Knopf, 1971.

Strogatz, Steven. *The Joy of X: A Guided Tour of Math, from One to Infinity*. Boston: Houghton Mifflin Harcourt, 2012.

Strogatz, Steven. *Sync: How Order Emerges from Chaos in the Universe, Nature, and Daily Life*. New York: Hypersion, 2003.

Susskind, Leonard. *The Cosmic Landscape: Strong Theory and the Illusion of Intelligent Design*. New York: Little, Brown and Company, 2006.

Susskind, Leonard and George Hrabovsky, *The Theoretical Minimum: What You Need to Know to Start Doing Physics*. New York: Basic Books, 2013.

Swift, Jonathan. *Gulliver's Travels and Other Writings*, Louis A. Landa, ed. Boston: Houghton Mifflin, 1960.

Szymborska, Wislawa. *View with a Grain of Sand: Selected Poems*. New York: Harcourt Brace, 1993.

Tattersall, Ian. *Becoming Human: Evolution and Human Uniqueness*. New York: Harcourt Brace, 1998.

Teller, Paul. "The Philosopher Problem" (634) in *The Rise of the Standard Model: Particle Physics in the 1960s and 1970s*, Lillian Hoddeson, et al, eds. Cambridge: Cambridge University Press, 1997.

Thomas, Andrew H. *Hidden in Plain Sight 2: The Equation of the Universe*. Kindle edition, 2013.

Thomas, Andrew H. *Hidden in Plain Sight 3: The Secret of Time*. Aggrieved Chipmunk Publications, 2014.

Thomas, Andrew H. *Hidden Plain Sight 7: The Fine Tuned Universe.* Aggrieved Chipmunk Publications, 2017.

Thomas, Keith. *Man and the Natural World: A History of the Modern Sensibility*. New York: Random House, 1983.

Thomas, Keith. *Religion and the Decline of Magic*. New York: Oxford University Press, 1971. Thomas, Lewis. *The Lives of a Cell*. New York: Viking Press, 1974.

Till, Charles E. and Yoon Il Chang. *Plentiful Energy: The Story of the Integral Fast Reactor*. San Bernadino, CA.: np, 2011.

Toliver, Harold. *Animate Illusions: Explorations in Narrative Structure*. Lincoln: University of Nebraska Press, 1974.

Tomita, K., "Periodically Forced Nonlinear Oscillators," in Arun V. Holden, ed. *Chaos*. Princeton: Princeton University Press, 1986.

Trefil, James, *The Nature of Science*. New York: Houghton Mifflin, 2003.

Turchin, Peter. *War and Peace and War: The Rise and Fall of Empires*. New York: Penguin, 2007.

Twain, Mark. *Extract from Captain Stormfield's Visit To Heaven*. New York: Oxford University Press, 1966.

Twain, Mark. *Is Shakespeare Dead?* New York: Oxford University Press, 1996.

Twain, Mark. *Sketches New and Old*. New York: Oxford University Press, 1996.

Unger, Roberto Mangbeira and Lee Smolin. *The Singular Universe and the Reality of Time: A Proposal in Natural Philosophy.* Cambridge: Cambridge University Press, 2015.

Union of Concerned Scientists. *The Climate Deception Dossiers*. ucsusa.org/sites/default/files/attach/2015/07/The-Climate-Deception-Dossiers.pdf

Union of Concerned Scientists. *Nuclear Power In a Warming World: Assessing the Risks* (2007).

Vaihinger, Hans. *The Philosophy of 'As If'*. Trans. C. K. Ogden. London: Routledge and Kegan Paul, 1935.

Van Diver, Bradford B. *Imprints of Time: The Art of Geology*. Missoula: Mountain Press, 1988.

Van Flandern, Tom. *Dark Matter, Missing Planets, and New Comets*. Berkeley: North Atlantic Books, 1993.

Veblen, Thorstein. *The Theory of the Leisure Class*. New York: Macmillan, 1899. Veneziano, "The Myth of the Beginning of Time" in *A Question of Time: The Ultimate Paradox*. New York: Scientific American, 2012.

Vickers, Brian, ed. *English Science: Bacon to Newton*. New York: Cambridge University Press, 1987.

Victoria, Brian A. "Holy War: Toward a Holistic Understanding," The Library of Social Science. *Ideologies of War* series. Http://www.libraryofsocialscience.com/ideologies-/docs/victoria_holy_war.html.

Von Humboldt, Alexander. *Works*. The Perfect Library: Kindle edition. Includes *Cosmos, Personal Narrative of Travels, America*.

Wallace, David Foster. *Everything and More: A Compact History of 00*. New York: Norton, 2003.

Wegener, Alfred. *The Origin of Continents and Oceans*, John Biram, trans. New York: Dover Publications, 1966. First German edition, 1912, English translation 1915.

Weiner, Jonathan. *The Beak of the Finch: A Story of Evolution in Our Times*. New York: Alfred A. Knopf., 1994.

Weinberg. Steven. *To Explain the World: The Discovery of Modern Science*. New York: Harper, 2015.

Weisman, Alan. *Countdown: Our Last, Best Hope for a Future on Earth?* New York: Little, Brown and Company.

Weizsäcker, Ernst von and Karlson Hargroves, Michael H. Smith, Cheryl Desha, and Peter Stasinopoulos. *Factor Five: Transforming the Global Economy through 80% Improvements in Resource Productivity*. London: Earthscan, 2009.

Wells, Spencer. *The Journey of Man: A Genetic Odyssey*. New York: Random House, 2003.

White, Michael. *Rivals: Conflict As the Fuel of Science*. London: Vintage, 2001. White, Patrick. *The Tree of Man*. New York: Penguin Books, 1961.

Whitson, Tom D., ed. *Weeds of the West*. University of Wyoming, 1996. 5th ed. Whittle, Mark. *Cosmology: The History and Nature of Our Universe*. Chantilly, Virginia: The Great Courses, 2008.

Wilson, Edward O. *Consilience: The Unity of Knowledge*. New York: Random House, 1998.

Wisdom, J., "Chaotic Behaviour in the Solar System," in Michael V. Berry, et al, eds, *Dynamical Chaos: Proceedings of the Royal Society of London*. Princeton: Princeton University Press,1987.

Wittgenstein, Ludwig. *The Philosophical Investigations*, George Pitcher, ed. New York: Doubleday & Company, 1966.

Wittgenstein, Ludwig. *Tractatus Logico-Philosophicus*, C.K. Ogden, trans. Mineola, New York: Dover publications, 1999 (first published 1922).

Wolff, Milo. *Exploring the Physics of the Unknown Universe: An Adventurer's Guide*. Manhattan Beach, California: Technotran Press, 1994.

Wolff, Milo. *Schrödinger's Universe: Einstein, Waves & the Origin of the Natural Laws*. Parker, Colorado: Outskirts Press, 2008.

Wolfson, Richard. "Physics and Our Universe: How It All Works" (2017), video lecture series.

Wolfson, Richard. *Simply Einstein: Relativity Demystified*. New York: W. W. Norton and Company, 2003.

Wolgar, Steve. *Science: The Very Idea*. New York: Tavistock Publication, 1988. Wolpert, Lewis. *The Unnatural Nature of Science*. Cambridge: Harvard University Press, 1992.

Wong, Kate, "Lucy's Baby," in *Becoming Human: Our Past, Present and Future*. New York: Scientific American, 2013.

Wood, Robert Muir. *The Dark Side of Earth*. London: Allen and Unwin, 1985. Wordsworth, William. *Poetical Works*. New York: Oxford University Press, 1973. World Bank. Open Data Catalogue: http://datacatalog.worldbank.org/.

Worldwatch Institute. *Is Sustainability Still Possible?* Washington, D.C., 2013.

Wulf, Andrea. *The Invention of Nature: Alexander Von Humboldt's New World*. New York: Alfred A. Knopf, 2017.

Yeats, W. B. *The Collected Works*, vol. 1, *The Poems*, Richard J. Finneran, ed. New York: Macmillan Publishing Company, 1989.

Yee, Jeff. *The Particles of the Universe*. Kindle edition, 2013.

Zimbardo, Philip. *The Lucifer Effect: Understanding How Good People Turn Evil*. New York: Random House, 2007.

Zukav, Gary. *The Dancing Wu Li Masters: An Overview of the New Physics.* New York: Bantam Books, 1980.